SpringerBriefs in Plant Science

SpringerBriefs present concise summaries of cutting-edge research and practical applications across a wide spectrum of fields. Featuring compact volumes of 50 to 125 pages, the series covers a range of content from professional to academic. Typical topics might include:

- A timely report of state-of-the art analytical techniques
- A bridge between new research results, as published in journal articles, and a contextual literature review
- A snapshot of a hot or emerging topic
- An in-depth case study or clinical example
- A presentation of core concepts that students must understand in order to make independent contributions

SpringerBriefs in Plant Sciences showcase emerging theory, original research, review material and practical application in plant genetics and genomics, agronomy, forestry, plant breeding and biotechnology, botany, and related fields, from a global author community. Briefs are characterized by fast, global electronic dissemination, standard publishing contracts, standardized manuscript preparation and formatting guidelines, and expedited production schedules.

More information about this series at http://www.springer.com/series/10080

Vladimir L. Gavrikov

Stem Surface Area
in Modeling of Forest Stands

 Springer

Vladimir L. Gavrikov
Siberian Federal University
Krasnoyarsk, Russia

ISSN 2192-1229 ISSN 2192-1210 (electronic)
SpringerBriefs in Plant Science
ISBN 978-3-319-52448-1 ISBN 978-3-319-52449-8 (eBook)
DOI 10.1007/978-3-319-52449-8

Library of Congress Control Number: 2017932912

Printed on acid-free paper

This Springer imprint is published by Springer Nature
The registered company is Springer International Publishing AG
The registered company address is: Gewerbestrasse 11, 6330 Cham, Switzerland

*To my wife Lena and
daughters Lera and Lisa*

Preface

To use a biogeographical analogy, scientific books may be divided into two categories. The first come under the motto, "All species in a limited area." Forest mensuration books often collect all methods, models, forest stand structures, and approaches belonging to this particular scientific area.

By contrast, books of the second category concentrate on "One species but everywhere." A good example of such an approach is given by dendrochronology. The focus of the branch of learning is practically limited by the width of a tree ring and its relation to the climate. But this focus spreads to all woody plants in all conditions and even in ancient epochs.

This book belongs rather to the second category.

In biology, the concept of a self-reproducing biounit—the cell—is one of the pillars of the science. All the macroscopically observed body functions such as respiration, movement, growth, and the like are overwhelmingly respiration and growth of the cells with the background of the macroscopic picture. Quantitative biological research should therefore imply consideration of the number of cells or their total mass, that is, biomass. Of course, a study of an animal population is neither possible nor necessary in terms of cell number. It is enough mainly to know the quantity of individuals or their biomass, because bodies of animals are metabolically active throughout the entire volume and thus one can expect the body biomass to be proportional to the number of active cells.

In the course of a growing season, bodies of herbaceous plants are also living cells. But woody plants, especially large tree species, have quite a different architecture. Although seedlings—like herbs—are metabolically active in the whole body volume they also accumulate dead mass inside stems as they age and living cells remain largely in a thin sheath covering the stems. Of course, stems are not the only home of living cells in trees. Foliage, buds, and roots are other organs containing living cells.

But this book is focused on tree stems as popular objects of study in forest science over many years. I tried to collect all available information on stem surface area scattered in numerous books and articles having, however, little hope of providing a complete guide because the accumulated storage and current flow of scientific information is indeed large.

Before going directly to the book content I would like to acknowledge the role of those persons whose scientific contributions provided especially important stimuli for my own studies.

First, one person is Professor Akio Inoue from the Prefectural University of Kumamoto, Japan. His publications and ideas evoked my interest in the topic and urged me to do my own research.

Second, are the researchers who published untreated data on forest stand structure and development, which is a rare case. Not publishing initial data may be psychologically quite understandable but it surely hampers the development of science.

I am feeling therefore deep gratitude to J.F. Bell, R.O. Curtis, J.E. King, and D.D. Marshall from the Pacific NW Research Station, US Forest Service. Also, Professor Dr. Vladimir A. Usoltsev collected and published a large database on forest stand structure for many species over Eurasia. These data were intensively used in the book and eventually made the writing of the book possible.

I would also like to acknowledge with gratitude detailed suggestions made by an anonymous reviewer during the manuscript preparation, which served to improve the manuscript.

Krasnoyarsk, Russia   Vladimir L. Gavrikov
July 2016

Contents

Chapter 1
Stem Surface Area as Subject of Study

Abstract In this chapter, a brief introductory review of tree biology is given. Such a review is meant to preface further analysis and to explain the approach implemented in the book. Especially, a look at how trees basically grow would be useful to clarify the way of subsequent analysis. An analysis of forest structure and growth, in particular its mathematical modeling, is intended to simplify the picture of trees and forest, that is, to consider some of their traits to be more important than others are. An account of biological fundamentals of trees may be rather helpful in doing the work of simplification as it helps to recollect important features of trees that make them quite peculiar biological organisms.

Keywords Plant growth • Primary growth • Secondary growth • Stem surface area • Trees

1.1 Primary and Secondary Growth: Version by Trees

Growth is one of the key notions as long as one considers the life cycle of plants. The notion of growth would, however, be dependent on the particular magnification at which it is viewed, that is, at which level of biological organization it is studied. Plant physiologists and anatomists would consider growth differently from ecosystem ecologists, for example.

In introductory remarks to the book *Biochemistry and Physiology of Plant Hormones*, Moore [19, p. 1] gave the following view of growth:

> "Growth" is defined as an irreversible increase in size that is commonly, but not necessarily ..., accompanied by an increase in dry weight and in the amount of protoplasm. Alternatively, it may be viewed as an increase in volume or in length of a plant or plant part. ...growth includes cell division as well as cell enlargement.

© The Author(s) 2017
V.L. Gavrikov, *Stem Surface Area in Modeling of Forest Stands*,
SpringerBriefs in Plant Science, DOI 10.1007/978-3-319-52449-8_1

While working at the level of tissues it is important to "see" cells, to explicitly take them into account, which is reflected in the desire of authors to distinguish between "growth" and "cellular differentiation" [31, 45]. Looking at the growth of a tree as a whole organism, not to speak about a community of trees, implies much less detail so that cells are no longer discernible. From the viewpoint of forest mensuration science, taking into account the cellular structure of trees would be redundant. Many studies cited and described in the subsequent chapters also exploit the approach of less detailed descriptions according to which growth is the increase of the linear, volumetric, or mass amount of an organism.

Being methodologically quite a sound one the approach still requires at least an implicit comprehension of a basic fact: it is a collection of *living* cells that constitutes the core essence of an organism, trees included. Neglecting the cellular nature of organisms may lead to a not quite optimal choice of morphological parameters when planning a study or model, whereas such a simplifying choice always has to be done due to a generally large amount of details describing an organism's growth and functioning. It is essential that the chosen morphological parameters have a close relationship to mass and form measures of that cell collection.

It is common knowledge that most vascular land plants, with minor exceptions, possess two kinds of growth: *primary* and *secondary* growth. The distinction between the two fundamental processes has been well studied for a long time and has become a standard content of botanical textbooks (see, e.g., [24, 28]).

Primary growth has been inherited by contemporary woody and herbaceous plants from ancient multicellular plant organisms that invaded the land. Primary growth is put into effect by apical meristems and results in elongation of the plant body along a growth axis. With respect to a tree, primary growth is associated with elongation of its shoots, twigs, leaves, and so on. Growth of the apical shoot at the very top of a tree provides a linear increase of its main stem and lateral shoots give rise to branches and start their further elongation.

The apical meristem, or apex, varies in its structure among plant species. Most woody plants have a layered apex structure consisting of an outer layer (tunica) and an underlying tissue of central mother cells, or corpus. The tunica cells divide anticlinally (at right angles to the surface of the apex), which provides an increase of the apex surface. The mother cells divide in random directions giving rise to an increase of the apical cone volume and eventually of length and primary thickness of the shoot.

Among both tunica cells and mother cells there are particular cells called initials. The sparse group of the initial cells, in an extreme case a single cell, placed on the tip of the apical cone is the only source of cells for all visible vegetative growth. Romberger et al. [26] gave a detailed histological picture of apical cone dynamics. In order to express the process in words, the authors suggested that readers imagine a cell flux originating from the apex and then downwards along the growth axis. If one puts an observer at the level of the uppermost initial it would seem to the observer that the cells flow out of the initial and stream downwards into the shoot where they differentiate, cease to divide, and, sooner or later, die. Only the cell at the front edge of the apical cone stays ever young and functions to produce new cells.

The mother cells divide relatively sparsely, whereas their descendants divide actively to produce primary tissues. Frequency of mitosis in the apical zone in many plants has been found to be at least fivefold lower than in the cells of underlying tissues. These descendant cells actively divide and grow achieving large sizes, up to 30,000 times larger than their ancestors, meristem initials [5]. Slow division and growth enable the longevity of initial cells. There is a wide notion that active mitosis and frequent DNA replication increase the risk of genetic information loss and degradation. A suppression of DNA synthesis lowers the risk and prolongs the cell cycle. An initial may have as few as only one or two divisions during seasonal growth [26]. It has been found in a popular hedge shrub, *Ligustrum ovalifolium*, that the initial cell divided only once in 12 days, during formation of three nodes. Also, the initial cell maintained its position as a center of growth for a very long period, during formation of 100 nodes [38].

The amazing properties of plant apical initials brought about a recent discussion in the scientific community of whether these founder cells are equivalent to pluripotent stem cells in animals, which resulted in the appearance of a concept of stem cells in plants. Inasmuch as the questions, although very interesting, are beyond the scope of this book, the reader can refer to a number of publications in which opinions on this matter have been summarized [30, 32, 46]. Briefly speaking, there are similarities as well as apparent differences between plant stem cells and animal stem cells but Laux [16] concluded that the similarities outweigh the differences. From his viewpoint, plants and animals appeared to evolve similar strategies independently to maintain undifferentiated cells that serve to cover the new cell requirements of the organism.

Primary growth results in primary tissues that include the most important tree plant organs such as buds and leaves. The buds possess their own primary growth; that is, primary growth creates other sources of primary growth. Inside the shoot, the primary cells differentiate into special organs usually denoted as vascular bundles. As has been extensively documented by plant anatomists ([24, 28, 49] and many others), a typical primary vascular bundle consists of the xylem and phloem that are produced by procambium, the meristem that itself stems from the apical meristematic cells.

In plant physiology, the question of which molecular signals mediate initiation of procambial activity has been widely explored (for a review of this issue, see Ye [49]). Auxin, the broadly known plant hormone, is transported from the apical zone downwards and a great deal of evidence has been found to substantiate that the hormone is the main factor of procambium initiation.

Obviously, the sites where procambial activity is first initiated determine the internal structure of plant stems because it is in these sites that the primary tissues (xylem and phloem) develop. By origin, the xylem and phloem have to be juxtaposed and divided by a thin layer of procambium. Figure 1.1 schematically gives an idea of the transverse section of a typical collateral primary vascular bundle.

Fig. 1.1 A schematic view of
a collateral vascular bundle,
transverse section:
Ph = phloem,
C = procambium, X = xylem,
GT = ground tissue

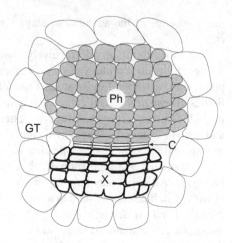

Functionally, the vascular bundles present transport pathways to deliver water up the stem (through the xylem) and to deliver photosynthesis products down the stem (through the phloem).

In stems of conventional trees, conifers and dicotyledons, the primary cells eventually differentiate and die, except those transformed into secondary meristem (cambium). It means that the primary growth itself is a rather fleeting event in these organisms; it starts and ends within a seasonal cycle of growth or between two states of dormancy. In the next cycle, it has to be initiated from the very beginning, from surviving initials. Such a habit of growth, rhythmic in time, is adequately reflected in space; the stems of trees are essentially modular from an architectural viewpoint.

A remarkable exclusion from the rule is palms, monocotyledon trees without secondary growth. Anatomical studies of palm stems [40–42] revealed that differentiated palm stem cells descending from primary apical meristems can be alive for centuries. In the whole bulk of palm stems, their volume is penetrated by numerous, literally thousands [28] of functioning vascular bundles scattered throughout the central ground tissue and all linked by multiple interconnections. All of the cell mass is metabolically active, hence living, because most of the cells show mitotic activity and undergo enlargement. In many respects, therefore, a palm "resembles the closed (unitary) growth of most animals rather than the open (modular) growth of most plants" [40, p. 9].

Secondary growth is the enlargement in plant stem thickness, that is, growth in the direction perpendicular to primary growth direction. The notation of the growth as "secondary" is linked to that it appears later than the primary growth, both phylo and ontogenetically. As a result of paleobotanical studies, secondary growth was found in the Middle Devonian plants and its appearance is dated back to approximately 380 million years ago. Interestingly, secondary growth seems to appear in a number of vascular plant clades that were not akin to each other [24]. The meristem of second growth is usually denoted as vascular cambium. Vascular cambium is present in extant gymnosperms and dicotyledon trees, as well as most herbaceous dicotyledons.

It should be noted that whereas secondary growth is considered to be a separate process from primary growth, from a molecular basis viewpoint there are definite similarities between the kinds of growth. Uggla et al. [43] have demonstrated that the same hormone, auxin, that regulates primary growth plays a role of positional signal in the vascular cambium of Scots pine (*Pinus sylvestris*). The radial width of the auxin concentration gradient was strongly correlated with the cambial growth rate and the largest concentrations of auxin were found in the fastest-growing cambia.

It has been widely supposed that evolutionarily new developmental processes often co-opt mechanisms and genes developed at earlier stages, which is true for secondary growth as well [8, 10]. Analyses of gene expressions showed that some genes regulating apical meristem functioning were found to be expressed in the vascular cambium. A comparison of an herbaceous *Arabidopsis* and woody *Populus* demonstrated that several genes may be involved in the genetic regulation in both meristems, which suggests that phylogenetically there was a conservation of regulatory mechanisms; that is, vascular cambium could inherit some of the mechanisms from the primary meristem [3]. The kinship of the basal molecular mechanisms in conventional trees and herbs with secondary growth allowed researchers to infer that "secondary growth is a measure of degree, rather than a trait that is present or absent" [9, p. 210].

Ontogenetically, vascular cambium becomes active at a short distance behind the apex of the stem; it is first initiated within vascular bundles between the xylem and phloem [28]. In most herbs with secondary growth the vascular bundles are dispersed in the stem volume as well-defined morphological structures separated by areas of ground tissue. Some herbs, however, form a distinct ring of vascular bundles. For example, in stems of alfalfa (*Medicado*) the ring consists of rather densely placed vascular bundles connected by stripes of interfascicular cambium. The morphological structure may therefore appear as a continuous layer of vascular cambium. Still the cambium layer does not function as a comprehensive whole in the herb. Within the alfalfa vascular bundles, the cambium shows small growth producing small quantities of secondary tissues, whereas interfascicular cambium produces only a small amount of sclerenchyma cells to the xylem side [24].

Only woody plants implement the structural layout of vascular cambium to an extreme degree. First initiated in vascular bundles, vascular cambium eventually extends between the bundles and forms a complete vascular cylinder all over the plant body. Woody species are predominantly characterized by *woody architecture*, which basically is a cylinder of dense stiff wood [27]. The cylinder formation process may be schematically represented as in Fig. 1.2.

As has been very well documented by plant anatomy, the functioning of vascular cambium is a many-stage process. Very similarly to growth of the apex, cambium initials produce mother cells—which are immature xylem cells—that continue to divide and only later produce mature xylem cells differentiating into vessels or tracheids [4]. The vascular cambium cells together with dividing immature

Fig. 1.2 A schematic view of woody architecture appearance through initiation of vascular cambium. (**a**) Primary growth, (**b**) formation of unbroken cylinder of vascular cambium, (**c**) conventional secondary growth (a picture of second-year growth), vb = vascular bundles, ic = interfascicular cambium, vc = vascular cambium, sph = secondary phloem, sx = secondary xylem, ar = annual ring of first year of growth

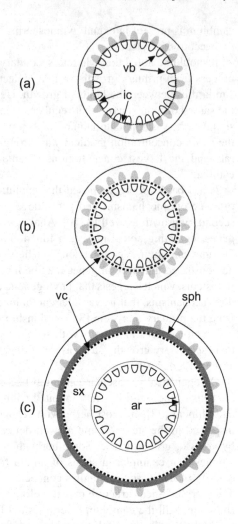

derivative cells are known to form a cambial zone. In some species, such as *Pinus strobus* [48], the cambial zone presents a large population of living cells.

This unbroken sheath of active meristematic cells covers every spot of the stem, except shoots undergoing primary growth. In woody plants with normal secondary growth, the vascular cambium generates secondary phloem at its outer edge; the secondary phloem cells eventually transform to periderm that serves as a barrier between the internal environment and the external medium.

The latter ability allows woody plants to satisfy a number of competing demands successfully. Among them are the necessities to grow larger (and locally adapt through growth), and to withstand an uncontrolled loss of water in a water-unsaturated atmosphere and still be able to perform CO_2/O_2 exchange with the atmosphere.

All the parts of vascular plants that appear in primary growth are covered by an outer layer of cuticle, a noncellular hydrophobic film. The cuticle plus epidermis provide both an effective barrier against dehydration and a gate for gas exchange through epidermal stomata. The architectural problem with the cuticle is that it is not renewable; enlargement of the plant primary organ or any injuries to it may lead to its fatal dysfunction. The problem is solved by implementation of phellogen, or cork cambium, another meristematic layer in the cortex. In stems of woody plants, the cuticle–epidermis–stomata complex is changed to phellem, or cork, that is interspersed with lenticels, areas regarded as paths for gas exchange, which provide effective protection and gas exchange under conditions of continuous secondary growth [17].

A look at a tree at the cellular level also provides understanding of tree longevity. A number of biological properties of trees contribute to a significant extension of the life cycle. Some of the properties are considered to be unique to trees and include (1) retention of meristematic activity after a period of dormancy, (2) replacement of damaged organs, (3) a sectored vascular system allowing a part of a tree to survive, and others [15]. It could be hence noted that in large part tree longevity is ensured by attributes of vascular cambium in that form in which it has been developed in trees.

Trees are plants but plants in general, even considering only higher plants, are so various that different parts of plant science developed their own foci of study and terminology. Short-lived plants such as annuals are very convenient objects of research because they can easily be grown in a laboratory and can be harvested as a whole body so their growth may be followed in great detail, from various size measurements to element contents. It has been found that the relation between measures is not constant as the growth of a plant progresses. Evans put forward an idea to focus on "ontogenetic changes in form and functioning, known generally as ontogenetic drift" [7, p. 18]. This term is seldom applied to trees. Instead, change in stem proportions is viewed in terms of self-similarity of stems and other tree organs as the growth progresses. Relationships between tree height, driven by primary growth, and stem thickness as a result of secondary growth were long ago found to be nonlinear, and may also be described in terms of ontogenetic drift. Explanations of why height growth experiences a slowdown have been suggested (see, e.g., hydraulic limitation hypothesis; [12]).

1.2 Evolution and Ecological Consequences of Secondary Growth in Trees

On the whole, secondary growth via bifacial cambium is regarded, along with the appearance of leaves and incorporation of lignin into cell walls, to be one of the major innovations during terrestrial plant evolution [3, 27].

For a structured and condensed picture of the evolution Bateman et al. [2] suggested viewing all the approximately 1800 million years of plant evolutionary development as four successive phases. The biochemical phase was the establishment of fundamental biochemical pathways related to respiration and photosynthesis. The anatomical phase characterized the appearance of most of the tissue types currently present in the extant plants. At the morphological phase, various arrangements of tissue types were evolutionarily tested; the maximum complexity of morphological forms as well as maximum body sizes were achieved. At last, the behavioral phase showed an extensive development of the coevolution of plants with micorrhizal or pathogenic fungi, with animal pollinators, dispersers, and the like [2].

The culmination of the morphological phase occurred in the Devonian when secondary growth appeared [2]. Erstwhile secondary xylem was not, however, a solution of the mechanical support task and did not allow trees to form large self-supporting stems. Moreover, it may have reduced the stiffness of the stem as secondary growth ruptured the outer sheath of the primary stiff material [27]. Probably the early xylem chiefly influenced water conductance [2].

Polymerization of monolignols in the cell walls is deemed to be the innovation that solved the problem of mechanical support because it has strengthened wood as the building material [3]. As a result of xylem lignification, large stiff stems of trees became a widespread growth form that allowed trees to bring the assimilation organs high up and compete with evolutionarily more advanced monocotyledons.

Tallness of the plant stem in general is considered to be a survival adaptive strategy in that sense that "[I]ncreasing plant size is adaptive on land and . . . tree-like morphologies bearing lateral planated branching systems or foliage leaves occupy adaptive peaks" [20, p. 278]. The strategy, however, has many problems because of the tall stem (a description of the trade-offs of tree growth form may be found in the review by Petit and Hampe [23]). In recent years, size-dependent features of tree growth and survival became a focus of studies. Zhang et al. [51] considered large physiological costs that tall neotropical savanna trees may bear. The costs include a decrease in stem-specific conductivity, vulnerability of leaves to embolism, and a more severe water deficit, among others. Nevertheless, all these disadvantages may be outweighed by advantages in a wetter climate. Thomas et al. [39] estimated that average mortality in the tallest lowland neotropical rain forest trees was twofold lower (1.2% per year) than was average in the landscape (2.7% per year).

Particularly, the availability of an abundant water supply may restrict distribution of tree-like growth forms. As discovered several decades ago, internal water deficit caused by excess transpiration or poor uptake from soil leads to inhibition of growth [13, 14], which inevitably minimizes the tall stem advantage. Even under ample soil moisture there are definite limitations to height growth. Koch et al. [12] studied *Sequoia sempervirens* in a wet temperate climate. Water stress and path length resistance were measured and found to increase with height that may limit leaf expansion and photosynthesis and hence further height growth. The authors have also demonstrated that twigs and needles attached to them dramatically decreased with height position in the crown of the species. Ryan et al. [29] pointed out, however, that hydraulic limitations occurring with tree height increase were common but not universal. Based on an extended literature review they suggested that tall trees differed from smaller and younger ones physiologically and that functional performance such as stomatal conductance and photosynthesis were often lower in taller trees but not always observed.

Still, water availability should be a major factor of forests' productivity. It has been recently suggested that water supply alone explained 60% of climate effect on tree growth in neotropical forests [44].

One feature of the discussions about height growth limitations in trees may be noted in the light of the topic of the chapter. The focus is largely on problems occurring in leaves. The leaves are meanwhile known to be the products of primary growth of buds that also determine length of annual shoots and hence height growth. It can be agreed that leaves are primary in function because they are producers of carbohydrates but buds (apical cones) are primary in determining how many leaves will appear as well as their size. Therefore, growth limitations taking place in buds might also be a focus in the framework of growth research. It is rather seldom that meristems are specifically taken into consideration in growth models. A recent example is given in the research by Hayat et al. [11] who introduced an explicit consideration of amounts of primary and secondary meristems in a model of tree growth. The method of modeling was due to the desire to consider very important limitations that take place in tree growth, namely, that there is a maximal rate at which individual cells can grow. In this respect, the authors distinguished their model from many other models of growth that were mostly photosynthesis driven.

At present, as much as 15–25% of the total amount of vascular plant species are tree species [23]; it has been known for a long time that the figures may vary among continents reaching 70% in Australia, for example [37]. Evolutionarily, this share of tree species is still a rather dynamic issue because tree growth habit has evolved many times. As mentioned above, achievements in the analysis of molecular mechanisms in both woody and herbaceous plants with secondary growth have shown that genes regulating apical shoot growth may have been co-opted by phylogenetically younger vascular cambium. The inferences about the overlapping molecular mechanisms in the two types of meristems may therefore explain why secondary growth could be gained and lost easily during the speciation in at least eudicotyledons [9, 10].

The appearance of large perennial stems of trees also had the natural consequence of the development of multistoried forests on the Earth's surface. The forests created a particular physical and chemical environment with a new microclimate that had a feedback effect on themselves as well as on a large amount of organisms living in forests and interacting with trees. Arborescence, seed habit, and the ability of cambium to stop and renew growth allowed land plants to colonize large areas of the planet, which actually created the contemporary look of the Earth biosphere. Land vegetation, forests included, may have substantially increased chemical weathering and hence rich nutrient fluxes through rivers to coastal seawaters promoting algal blooms, anoxia of bottomwaters, and intensive sedimentation of organic carbon. These teleconnections between terrestrial and marine ecosystems might have had actual global effects on the life of the planet including the sinking of CO_2 and subsequent cooling of the atmosphere [1].

The conspicuous longevity of trees led to the microclimate of woodlands being rather prolonged in time allowing the evolution of a very specific forest biota of smaller plants and animals. From the viewpoint of an annual plant or small animal, trees can live almost eternally. Longevity of the oldest Earth trees may amount to thousands of years [34, 36].

The volumetric size of the tree stem, that is, its volume, is an important characteristic of a tree. It is the mass of lignified dead xylem inside the stem that stiffens it and allows it to dominate three-dimensional space high up from the ground. On the other hand, the tree as an organism does not interact with the environment directly by volume. The interaction is largely through surfaces of a tree, both as a biological and a physical body.

As a physical body, the tree stem surface modifies a number of energy and substance flows. For instance, the surface is supposed to absorb shortwave radiation coming from the sun and to radiate longwave energy into the surrounding environment [33].

In most forest hydrological studies, stemflow has usually been paid significant attention [18]; under stemflow the part of precipitation water flowing down the woody plant stem surface is understood. Obviously, the stemflow plays a minor role in dense tropical forests with highly developed canopies that are the main agents of precipitation interception in these sites [6]. Under other conditions, especially arid ones, a redistribution of precipitation by stem surfaces may, however, have larger effects. Shrub-like *Eucalyptus* in Western Australia intercepted by canopy only 15% of water that was then lost into the atmosphere but 25% ran down the stem, which had an observable consequence for the species survival. The effect was that the water was routed via stemflow directly to roots and then along them deeper into the soil, which contributed to the subsoil water pool used during dry times [21]. Moreover, it has been shown that the water running down the stems of arid *Larrea* bushes had significantly higher concentrations of nutrient ions than water of bulk precipitation, which presumably caused a "fertile island" effect on the soil under the bushes [47].

In the course of trees' evolution, a large group of specialized insects has evolved that infest tree stems, the xylophage insects. Although they are called *xylo*phages their feeding behavior is primarily directed to the inner bark, that is, phloem which is a living and rich in digestible nutrients tissue as contrasted with lignified xylem. The larvae of most sawyer beetles of the *Monochamus* genus feed in the phloem and only later tunnel into the xylem (sapwood) to pupate (e.g., *Monochamus scutellatus* [22]). This means that these insects interact with trees as surfaces. It is the stem surface that constitutes their nutritive base, and traditionally forest pest entomologists use area units to describe density emergence of a pest per bark area ([25, 35, 50, 52] and others).

As follows from the above descriptions, conventional trees evolutionarily implemented an extreme growth form architecture. Many herbaceous plant species possess secondary growth but trees developed a certain spatial growth structure. The structure is often described as an unbroken cylinder of cambial cells; another visual metaphor for the structure is a cell sheath that covers practically all the tree body (except the tips of shoots and leaves).

Because stem-only vascular cambium cells and a thin layer of their immediate descendant cells are living in a strict sense they constitute to large extent the living biological body of tree. In other words, a tree is alive because it possesses the living mass of cells (cambial and apical), not because of its linear or volumetric dimensions. As long as the living cells are spatially organized as a superficial layer, the most appropriate metaphor for a tree may be a surface that is essentially two-dimensional and therefore is measured in area units.

A tree may be seen as a surface not only in an organismal sense; it interacts with the environment through its surface, first of all, and performs the gas exchange (respiration of stem). Also, the surface plays the roles of a food reserve for xylophages and of a living substrate for epiphyte flora.

Eventually, the stem surface is the base for production of fuel-wood for house-keeping or commercial timber for use in the economy. As Schreuder put it, the surface of stems of a forest stand is "a measure of the amount of stand in a physical sense" [33, p. 248].

In subsequent chapters, the surface of tree stems is at the center of analysis and modeling, mostly as the total surface area of a stand, which gives us the chance to interpret alterations of the stem surface as a reflection of competition among trees.

References

1. Algeo TJ, Scheckler SE (1998) Terrestrial-marine teleconnections in the devonian: links between the evolution of land plants, weathering processes, and marine anoxic events. Philos Trans R Soc B Biol Sci 353(1365):113–130
2. Bateman RM, Crane PR, DiMichele WA, Kenrick PR, Rowe NP, Speck T, Stein WE (1998) Early evolution of land plants: phylogeny, physiology, and ecology of the primary terrestrial radiation. Annu Rev Ecol Syst 29:263–292
3. Baucher M, El Jaziri M, Vandeputte O (2007) From primary to secondary growth: origin and development of the vascular system. J Exp Bot 58(13):3485–3501
4. Beck CB (2010) An introduction to plant structure and development: plant anatomy for the twenty-first century. Cambridge University Press, Cambridge
5. Cosgrove DJ (2005) Growth of the plant cell wall. Nat Rev Mol Cell Biol 6(11):850–861. doi:10.1038/nrm1746
6. Dietz J, Hölscher D, Leuschner C et al (2006) Rainfall partitioning in relation to forest structure in differently managed Montane forest stands in central Sulawesi, Indonesia. For Ecol Manag 237 1:170–178
7. Evans GC (1972) The quantitative analysis of plant growth, vol 1. University of California Press, Berkeley
8. Ganfornina MD, Sánchez D (1999) Generation of evolutionary novelty by functional shift. Bioessays 21(5):432–439
9. Groover AT (2005) What genes make a tree a tree? Trends Plant Sci 10(5):210–214
10. Groover A, Robischon M (2006) Developmental mechanisms regulating secondary growth in woody plants. Curr Opin Plant Biol 9(1):55–58
11. Hayat A, Hacket-Pain AJ, Pretzsch H, Rademacher TT, Friend AD (2016) Modelling tree growth taking into account carbon source and sink limitations. bioRxiv 063594; doi:https://doi.org/10.1101/063594

12. Koch GW, Sillett SC, Jennings GM, Davis SD (2004) The limits to tree height. Nature 428(6985):851–854
13. Kozlowski TT (1971) Cambial growth, root growth, and reproductive growth, vol 2. Elsevier, New York
14. Kramer PJ (1962) The role of water in tree growth. In: Tree growth. Ronald Press, New York, pp 171–182
15. Lanner RM (2002) Why do trees live so long? Ageing Res Rev 1(4):653–671
16. Laux T (2003) The stem cell concept in plants: a matter of debate. Cell 113(3):281–283
17. Lendzian KJ (2006) Survival strategies of plants during secondary growth: barrier properties of phellems and lenticels towards water, oxygen, and carbon dioxide. J Exp Bot 57(11):2535–2546
18. Llorens P, Domingo F (2007) Rainfall partitioning by vegetation under mediterranean conditions. a review of studies in Europe. J Hydrol 335(1):37–54
19. Moore TC (1989) Introduction. In: Biochemistry and physiology of plant hormones. Springer, New York, pp 1–27
20. Niklas KJ (1997) The evolutionary biology of plants. University of Chicago Press, Chicago
21. Nulsen R, Bligh K, Baxter I, Solin E, Imrie D (1986) The fate of rainfall in a mallee and heath vegetated catchment in southern western Australia. Aust J Ecol 11(4):361–371
22. Peddle S, De Groot P, Smith S (2002) Oviposition behaviour and response of Monochamus scutellatus (Coleoptera: Cerambycidae) to conspecific eggs and larvae. Agric For Entomol 4(3):217–222
23. Petit RJ, Hampe A (2006) Some evolutionary consequences of being a tree. Annu Rev Ecol Evol Syst 37:187–214
24. Raven PH, Evert RF, Eichhorn SE (1992) Biology of plants, 5th edn. Worth Publishers, New York
25. Reagel PF, Smith MT, Hanks LM (2012) Effects of larval host diameter on body size, adult density, and parasitism of cerambycid beetles. Can Entomol 144(03):435–438
26. Romberger JA, Heinowicz Z, Hill JF (1993) Plant structure: function and development. A treatise on anatomy and vegetative development with special reference to woody plants. Springer, Berlin
27. Rowe N, Speck T (2005) Plant growth forms: an ecological and evolutionary perspective. New Phytol 166(1):61–72
28. Rudall PJ (2007) Anatomy of flowering plants: an introduction to structure and development. Cambridge University Press, Cambridge
29. Ryan MG, Phillips N, Bond BJ (2006) The hydraulic limitation hypothesis revisited. Plant Cell Environ 29(3):367–381
30. Sablowski R (2004) Plant and animal stem cells: conceptually similar, molecularly distinct? Trends Cell Biol 14(11):605–611
31. Salisbury F, Ross C (1985) Plant Physiol. Wadsworth, Belmont
32. Scheres B (2007) Stem-cell niches: nursery rhymes across kingdoms. Nat Rev Mol Cell Biol 8(5):345–354
33. Schreuder HT, Gregoire TG, Wood GB (1993) Sampling methods for multiresource forest inventory. Wiley, New York
34. Schweingruber FH (1996) Tree rings and environment. Dendroecology. Paul Haupt, Birmensdorf/Berne/Stuttgart/Vienna
35. Sekretenko O, Kovalev A, Soukhovolsky V (2015) Oviposition of the fir sawyer beetle (Monochamus urussovi Fisch.) on the tree stem: analysis using models of random point fields. J Sib Fed Univ Biol 1(8):45–55 (in Russian)
36. Sidorova OV, Naursbaev MM, Vaganov EA (2005) Longevity record holders among woody species. Lesnoe Hoz 5:23–24 (In Russian)
37. Sinnott EW (1916) Comparative rapidity of evolution in various plant types. Am Nat 50:466–478
38. Stewart R, Dermen H (1970) Determination of number and mitotic activity of shoot apical initial cells by analysis of mericlinal chimeras. Am J Bot 57:816–826

39. Thomas RQ, Kellner JR, Clark DB, Peart DR (2013) Low mortality in tall tropical trees. Ecology 94(4):920–929
40. Tomlinson PB (2006) The uniqueness of palms. Bot J Linn Soc 151(1):5–14
41. Tomlinson PB, Huggett BA (2012) Cell longevity and sustained primary growth in palm stems. Am J Bot 99(12):1891–1902
42. Tomlinson PB et al (1990) The structural biology of palms. Oxford University Press, Oxford
43. Uggla C, Mellerowicz EJ, Sundberg B (1998) Indole-3-acetic acid controls cambial growth in scots pine by positional signaling. Plant Physiol 117(1):113–121
44. Wagner F, Rossi V, Stahl C, Bonal D, Herault B (2012) Water availability is the main climate driver of neotropical tree growth. PloS One 7:e34074
45. Wareing PF, Phillips LDJ (1981) Growth and differentiation in plants. Pergamon Press, New York
46. Weigel D, Jürgens G (2002) Stem cells that make stems. Nature 415(6873):751–754
47. Whitford WG, Anderson J, Rice PM (1997) Stemflow contribution to the 'fertile island' effect in creosotebush, larrea tridentata. J Arid Environ 35(3):451–457
48. Wilson BF (1966) Mitotic activity in the cambial zone of pinus strobus. Am J Bot 53:364–372
49. Ye Z-H (2002) Vascular tissue differentiation and pattern formation in plants. Ann Rev Plant Biol 53(1):183–202
50. Zhang Q-H, Byers JA, Zhang X-D (1993) Influence of bark thickness, trunk diameter and height on reproduction of the longhorned beetle, Monochamus sutor (Col., Cerambycidae) in burned larch and pine. J Appl Ent 115:145–154
51. Zhang Y-J, Meinzer FC, Hao G-Y, Scholz FG, Bucci SJ, Takahashi FS, Villalobos-Vega R, Giraldo JP, Cao K-F, Hoffmann WA et al (2009) Size-dependent mortality in a Neotropical savanna tree: the role of height-related adjustments in hydraulic architecture and carbon allocation. Plant Cell Environ 32(10):1456–1466
52. Zhong H, Schowalter T (1989) Conifer bole utilization by wood-boring beetles in western oregon. Can J For Res 19(8):943–947

Chapter 2
Stem Surface Area: Measurement and Development

Abstract The ability to quantify an object of study is crucial for any scientific research. In our three-dimensional physical space, linear, that is, one-dimensional, sizes are measured quite easily. Three-dimensional sizes, volume and mass, may be measured with good precision as well, at least for objects of a human body order of magnitude. All these measurements can be, in fact, effectively done independently of the shape that the objects have, but not in the case of two-dimensional sizes such as surface area. For surface area, the shape is of primary importance: the more complicated it is, the harder it is to estimate surface area value. In forests, researchers face relatively large bodies—stems of trees—that more or less deviate from ideal shapes, which presents a methodological problem if one thinks of measurements in numerous replications necessary for statistical treatment.

Basically, there are only two ways to estimate a value of size. One is to measure it directly, when specialized equipment is at hand. Another, indirect, is to calculate it from other values that can be measured using appropriate formulas known beforehand. For example, mass may be measured directly if one has scales. Mass may also be estimated through known volume and mean density.

For surfaces, especially of three-dimensional shapes, only the latter method of calculation is practically available.

In this chapter, the issues of how stem surface area can be measured as well as how the area evolves with growth of a forest stand are considered.

Keywords Cone approximation • Douglas fir • Scots pine • Stem diameter • Stem length • Stem shape • Stem surface area

2.1 Stem Surface Area and Forest Mensuration

It would be an overstatement to say that the value of the stem surface area played an important role in forest mensuration in general. Still, some books paid sufficient attention to the stem dimension parameter [1, 17]. Also, some techniques included calculations of stem surfaces as a routine function (see, e.g., [6]).

© The Author(s) 2017
V.L. Gavrikov, *Stem Surface Area in Modeling of Forest Stands*,
SpringerBriefs in Plant Science, DOI 10.1007/978-3-319-52449-8_2

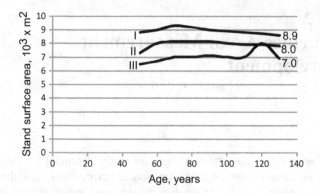

Fig. 2.1 Development of stem surface area in fully stocked Scots pine stands according to Anuchin [1]. Roman numbers on the *left* of curves are site classes. Arabic numbers on the *right* are means for correspondent curves calculated by Anuchin. The graphs were constructed by the author according to a table published by Anuchin [1, p. 456]

The interest in stem surface area has been linked to the common necessity in forest mensuration to estimate the volume increment of forest stands. Husch et al. [7] mentioned that stem surface area approximated cambial surface and therefore that surface on which the wood substance accumulates. All this made the stem surface area "useful in the estimation of tree and stand growth" [7, p. 95] and a group of mensuration methods was based on stem surface area.

Anuchin [1] commented that current volume increment Z_m was a function of cambium surface area S and radial increment Z_r : $Z_m = f(S, Z_r)$. Having this in mind, he undertook studies of stem surface areas in differently aged fully stocked stands of Scots pine. As a result, he received the estimations of stem surface areas for Scots pine stands of three bonitation[1] classes.

Figure 2.1 depicts development of stem surface area received by Anuchin [1]. He concluded that the stem surface area stays constant in Scots pine fully stocked stands from 50 years of age for site class I, from 60 years for site class II, and from 60 to 70 years for site class III. The corresponding means were estimated by Anuchin to be 8.9, 8.0, and 7.0 $10^3 \times m^2$ for site classes I, II, and III, correspondingly.

Anuchin also stated that the stem surface area should be proportional to the degree of stocking and, therefore, knowing the observed degree of stocking, one can reduce the fully stocked stand stem surface area for a given stand through multiplication by the observed degree of stocking. As a result, he suggested

---

[1]Russian system to reflect site quality which is based on bonitet classes. These classes are linked to the height of the stand at age 100 years. Class I of bonitet denotes best conditions; class V corresponds to worst conditions. In the following descriptions, "site classes" is used instead of "bonitet classes" as is more customary in the English literature. The numbers of the classes, however, correspond to the original Russian publications.

estimating the current volume increment in Scots pine stands from the stem surface area, mean thickness of tree rings, and degree of stocking through relationships of the sort: $Z_m^I = S_I tp = 8.9tp$, $Z_m^{II} = S_{II} tp = 8.0tp$, and $Z_m^{III} = S_{III} tp = 7.0tp$, where t stands for a mean thickness of tree rings and p for degree of stocking.

Concluding the description of the method, Anuchin [1] commented that, first, similar formulas may be derived for every species and site class and, second, that having at hand yield tables with numbers of trees and their mean dimensions one can calculate stem surface area. However, he made no suggestions in his book as to how to perform such calculations of stand stem surface area.

Regarding surface area of an individual stem, Husch et al. [7] gave a description of how the surface area of a body of revolution is calculated. Assuming that the form of a tree approximates a paraboloid and that the stem surface is received by revolving the equation

$$Y^2 = 0.066X, \tag{2.1}$$

about the X axis, one can calculate the surface area. In Eq. (2.1), which is a particular case of the Kunze formula, Y stands for the radius of the stem and X for the distance from the tip of the stem. From geometry it is known that such a surface of a body of revolution S_x may be found using the equation

$$S_x = 2\pi \int_0^{X_{max}} Y \sqrt{1 + \left(\frac{dY}{dX}\right)^2} \, dX. \tag{2.2}$$

Mathematically, the term dY/dX says how fast the variable Y changes with a change of X. If the matter is an exact calculation of surface area of a definite stem then it would be better to take it into account. On the other hand, as is well known, stems of conventional trees are rather narrow shapes so that the change of Y may be vanishingly small. If, therefore, one studies an analytical model the term $\sqrt{1 + (dY/dX)^2}$ makes the computations too tedious, adding no important sense into the model. It would, therefore, be rational to concede that $dY/dX = 0$. The inadvertent error and the loss of exactness have the advantage that formulas become much more operable and transparent (see also page 34).

In general, however, as mentioned above, stem surface area as a research subject is far from being considered in all forest mensuration books. Many of them, both old [3] and modern [19] show no interest in the matter.

2.2   Measuring Stem Surface Area for Research

Measurements of stem surface area performed for research purposes often aim at precision of measurement, which involves labor-intensive and time-consuming techniques. Swank and Schreuder [18] undertook thorough research to estimate error terms for a few methods of sampling in a 10-year-old eastern white pine (*Pinus strobus*) forest stand. The measured parameters were surface areas of foliage, branches, and stems as well as oven-dry weight of these tree organs.

As a motivation for the research, Swank and Schreuder commented that "Quantity of tree surface area and tree biomass per unit area of land are inventory data needed to understand the flow of energy, nutrients, and water through forest ecosystem. Surface area, in contrast to biomass, has received little attention" [18, p. 91].

Swank and Schreuder [18] sampled whole trees and first determined the geometrical form of the stems through multiple measuring of diameters along the stems. Plotting of diameters against heights of the measurements revealed that a conic shape was the best approximation. It was therefore natural to use the formula for the right cone to compute stem surface area S as $S = \pi r \sqrt{r^2 + h^2}$, r and h being the stem radius at ground level and height, respectively. The authors also established regression relationships linking stem surface area to some easily measured values. In particular, stem surface area appeared to correlate tightly with basal area and dbh. The coefficient and intercept for an untransformed relationship with the basal area were 31.29 and ≈ 1.09 ($R^2 = 0.984$), respectively. The relationship with dbh was log-transformed, with the coefficient value being 1.663 and intercept value being ≈ 1.38($R^2 = 0.994$). As a result, they found that foliage, branches, and stems of eastern white pine had 5.3, 0.76, and 0.13 ha surface area per ha of land surface.

Similar research, although with other purposes, was conducted by Pokorný and Tomášková [16]. In a detailed study of a 20-year-old Norway spruce (*Picea abies*) forest stand the authors tried to establish correlations between many stand measures important for ecophysiological studies and growth or carbon uptake models. Another motivation was to find relations between easily measured dendrometric parameters and those that were not readily measurable. Among the studied parameters were mass/volume and surface areas of various aboveground tree organs, from foliage and current growth shoots to the stem. Stem volume and its surface area were obtained through diameter measurements in the middle of 1 m long sections of stems, with the shape of the sections being assumed to be cylindrical. Specifically regarding stem surface area, the authors established that the area (StA, m^2) related to dbh (D, cm) and height (H, m) through nonlinear functions with high correlation strength: $StA = 0.0445D^{1.4696}$ ($R^2 = 0.92$) and $StA = 0.0279H^{1.8454}$ ($R^2 = 0.89$).

It is clear that the shape of the stem may be of importance in estimating the stem surface area. For many years it has been known that large trees may have different shapes in different parts of their stems. For example, Grosenbaugh [6] commented that trees tend to follow a neiloid shape closer to the butt. Further to the tip, the stem is a conoid and finally a paraboloid in the upper stem portions.

Such a deeply detailed elaboration is useful when measuring individual stems but raises considerable obstacles for modeling. For numerously replicated measurements as well as for mathematical analysis, it would be convenient to relate the factual stem surface area to the surface area of an ideal geometrical shape. Because conoid is in fact a part of a real stem shape, this figure may be a good candidate for the relation.

Stem analysis is a routine method of research in forest sciences. Briefly, stem analysis data present diameter measurements at regular intervals along the stem and across the stem. The measurements across the stem provide diameter values of the tree when it was younger, which is recorded in tree rings. The heights of the past stems inside the tree are also measured. Based on the data a researcher can reconstruct sizes of all the stems inside the tree and hence the growth of the tree. The stem surface area may also be well computed.

Such computations were performed for data on stem analysis of a 227-year-old Scots pine tree. For differently aged reconstructed stems, stem surface area was computed through two methods. First, the surface area was computed section by section assuming that each section of the analyzed stem was a truncated cone. Then all the obtained values were summed up to give a measured value for the entire tree ($S_{measured}$). Independently, the base diameter of a stem and its height were used to compute the surface area of a correspondent ideal cone (S_{conic}). Finally, the relation of $S_{conic}/S_{measured}$ was followed through the available range of ages (Fig. 2.2).

As can be seen in Fig. 2.2, the ratio $S_{conic}/S_{measured}$ is close to 1 for younger ages (cf. young eastern white pine stand in Swank and Schreuder [18]). Then the

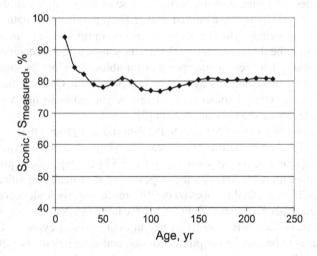

Fig. 2.2 Relation of measured stem surface area for a 227-year-old Scots pine to surface area computed for corresponding cone figures. $S_{measured}$ = stem surface area computed for each section of the tree and then summed up. S_{conic} = surface area of a cone of base diameter and height equal to the values in the tree. The stem surface area was computed from stem analysis data provided by Prof. V.V. Kuzmichev. The data are given in the Appendix, Tables A.19, A.20, and A.21

ratio decreases quite quickly until about 40 years of age, and then, as the tree ages, oscillates around 0.79–0.86. It may be suggested therefore that after a definite age of trees their stem surface area keeps in approximately constant relation to the surface area of the corresponding cone. The observation has been used in modeling of total stem surface development in Sect. 3.1.2.

Inoue published a series of works that covered various questions of measurement as well as development of stem surface area [8–11].

Application of a regression equation is a widely used method to estimate stem surface area. Thus Inoue [8] undertook a study aimed at collecting information on relationships between stem surface area and other stem dimensions, as well as uncovering the advantages and disadvantages of published equations and variations of the coefficients among species. The author used field data from forest stands of two species, Japanese cedar (*Cryptomeria japonica*) and Japanese cypress (*Chamaecyparis obtusa*). Fifty trees of each of these species were felled and nine diameters of the stems were measured at equal intervals along the stems. Then the data were fitted by third-order polynomial function, the distance from the tip of the stem being independent (X) and diameter dependent (Y) variables. To get the stem surface area, Inoue used an integration formula such as (2.2) assuming $dY/dX = 0$.

Inoue [8] also gave a brief critical review of equations suggested by various authors to predict stem surface area. Some of them used diameter as an independent variable; others used both diameter and height of trees. Inoue discarded linear equations of the sort $S = \alpha X + \beta$ (S is the stem surface area, α and β are coefficients, and X is an independent variable, dbh, or basal area) arguing that the equations must go through the origin as the surface area must be zero at zero dbh. Equations like $S = \alpha D^\beta$ (D denotes dbh) were also discarded because of inconsistency of dimensions except when $\beta = 2$. Inoue admitted that a three-parameter equation such as $S = \alpha D^\beta H^\gamma$ (H is stem height; γ is a coefficient) may provide the highest accuracy but he found that the determination of the coefficients may be inadequate because of multicollinearity between the independent variables, therefore the equation was discarded. Finally, two equations were taken for comparison. One uses basal area as $S = \alpha G$ (G is basal area). Another uses dbh and height and looks like $S = \beta DH$ as earlier suggested by Lexen [14] and Carron [2].

Through fitting of equation $S = \alpha G$ to the data Inoue [8] found that stem surface area is relatively good related to basal area. The slopes α were determined to be 184.216 for Japanese cedar ($R^2 = 0.67$) and 156.878 for Japanese cypress ($R^2 = 0.873$), with the difference between the slopes being statistically significant. Inoue explained the difference in the slopes by the difference of the bole slenderness of the species studied. Japanese cedar was shown to have more slender stems than Japanese cypress; that is, cedar stems have larger height than stems of cypress of the same dbh. Compared to basal area the product of dbh and stem height, as expressed by equation $S = \beta DH$, showed a strong positive correlation with stem surface area [8]. The slope β for Japanese cedar was 1.937 ($R^2 = 0.902$) and the slope for Japanese cypress was 1.921 ($R^2 = 0.961$). As Inoue reported, no significant difference was found between the slopes for the two species. He explained this finding in such a way that the product of dbh and total tree height reflected the bole slenderness property and therefore removed its effect from the regression. Because the β slopes were insignificantly different for the studied species Inoue discussed how wide the slope values varied among conifers. He cited a work by Lexen [14] in which the slope was reported to be 1.72 for *Pinus ponderosa* trees and a work by Carron [2] in which a variation of the slope from 1.64 to 1.85 was found in *Pinus radiata* plantations. Higher slope values for the two Japanese species Inoue explained by a difference in measurement methods; that is, the cited authors used inside bark stem surface area whereas Inoue measured outside bark surface area. Taking into account the relationships between inside bark and outside bark surface areas Inoue concluded that the variation in the slope should be small among conifer species.

As a result, Inoue [8] suggested that formula

$$S = \beta D_m H_m N \tag{2.3}$$

may be a useful approximation for a stand surface area, where D_m and H_m are mean dbh and mean height, respectively, and N is the number of trees. The approach of stand surface area estimation through the product of mean dbh and mean height is used below in Chap. 3.

In another publication, Inoue [9] explored the question of whether form-factors for stem volume can help in determining stem surface area. Form-factors are a widely known and applied concept in forest mensuration science. Basically, if a tree stem has a diameter D and height H a form-factor for the stem is a coefficient that links the stem volume with the volume of a cylinder having the same diameter D and height H. Because stem form would likely influence the coefficient the form-factor is considered to be a measure of stem form. The usefulness of the form-factor concept has been shown in a number of studies cited by Inoue in which the normal form-factors for conifers, provided the diameters were measured at relative heights 0.7 and 0.5, were reported to be almost steady at the values of 0.7 and 1.0, respectively, and to be "independent of species, district, density control, and growing stage" [9, p. 289].

Inoue [9] suggested a new measure of stem form based on stem surface area. By analogy with the usual form-factor it can be expressed as

$$\frac{S}{\pi D_i H} = \kappa_i, \tag{2.4}$$

where S is the stem surface area, D_i is the diameter at a relative height i, and κ_i is the form-factor. To differentiate the new measure from traditionally used form-factors for stem volume Inoue called κ_i "a form-factor for stem surface area."

If one could get universal values for κ_i they might be used to estimate stem surface area with the same approach as is done with form-factors for stem volume. The form-factors for stem volume represent, however, a very concentrated experience of forest science, a result of decades of hard work by forest scientists who collected data from various species, geographical locations, stand densities, and so on. Theoretically, if one would like to use the concept of form-factors for stem surface area the researcher also has to collect all the data. Inoue [9] put forward an idea that form-factors for stem volume and form-factors for stem surface area are both measures of stem form and if it were possible to link them then it would also be possible to use all the large experience accumulated on form-factors for stem volume without repeating the labor-intensive measurements.

As a starting point of the analysis, Inoue [9] took the well-known Kunze equation

$$r^2 = mh^n,$$ (2.5)

where r is stem radius, h is stem length from the tip, and m and n are parameters.

Through integration of formula (2.5) from 0 to h, Inoue [9] received expressions for volume V and surface area S of the stem as a body of revolution:

$$V = \frac{m\pi}{n+1} h^{n+1}, \quad S = \frac{4\sqrt{m\pi}}{n+2} h^{n/2+1}.$$ (2.6)

From (2.6), formulas for form-factors for stem volume λ and surface area κ were expressed by Inoue as

$$\lambda_i = \frac{V}{\frac{\pi}{4}D_i^2 h} = \frac{4mh^n}{D_i^2(n+1)}, \quad \kappa_i = \frac{S}{\pi D_i h} = \frac{4\sqrt{m}h^{n/2}}{D_i(n+2)}.$$ (2.7)

Finally, eliminating h and D_i from (2.7) Inoue got an elegant expression linking form-factors for stem volume and surface area:

$$\kappa_i = k\lambda_i^{0.5},$$ (2.8)

where k is a coefficient equal to $\dfrac{2(n+1)^{0.5}}{n+2}$. The result of the theoretical analysis by Inoue—expression (2.8)—says that the form-factor for stem surface area should be proportional to the square root of the form-factor for stem volume. If it is so, then form-factors for stem surface area may be estimated from known universal values of form-factors for stem volume.

In order to prove expression (2.8) Inoue [9] used data on Japanese cedar and Japanese cypress. Applying a major axis regression technique to the power equation $\log \kappa_i = \alpha + \beta \log \lambda_i$ he found that coefficient β was 0.505 for cedar trees and 0.501 for cypress trees, with the values not significantly different from 0.5 for both species.

Thus Inoue [9] was able to apply expression (2.8) to the data directly to get the coefficient k linking the form-factor for the stem surface area with that for stem volume. It was found that the resulting equations were $\kappa_i = 0.873\lambda_i^{0.5}$ ($R^2 = 0.989$) for cedar trees and $\kappa_i = 0.873\lambda_i^{0.5}$ ($R^2 = 0.996$) for cypress trees, with no significant difference between the coefficients for the species, which gave Inoue the opportunity to hypothesize their equality among the species.

As a result, Inoue [9] estimated values for normal form-factors for the stem surface area that he suggested to be universal ones for coniferous species. The values for normal form-factors for stem volume were reported in the literature to be 0.7 and 1.0 for relative heights $\lambda_{0.7}$ and $\lambda_{0.5}$, respectively. Through substitution of the values to $\kappa_i = 0.873\lambda_i^{0.5}$ Inoue received the required values as $\kappa_{0.7} = 0.730$ and $\kappa_{0.5} = 0.873$. He also showed that the breast height form factor for stem surface area κ_b depended on tree height decreasing sharply up to the height values of 10 m and for larger heights changing very slowly, approaching an asymptote of $\kappa_b = 0.605$.

2.3 Development of Stem Surface Area

Analogical to other forest stand measures of size, including stand volume or basal area, total stem surface area may be studied using both static and dynamic data. Static data are mostly sets of measurements performed at 1 year that are to be dynamically interpreted. An often-cited example of this kind of research is self-thinning laws. These laws represent allometric relationships between mean measures of tree size and stand density.

A self-thinning relationship may be interpreted to be the upper boundary in the space "mean size-stand density" depicting a maximum mean size at a given value of stand density. An allometric model was suggested by Inoue [10] to predict the slope of the boundary in terms of stem surface area versus stand density. Inoue used the concept of "biomass density" suggested earlier in the literature, [13] which means the mass per occupied space.

In order to formulate the model, Inoue [10] defined biomass density B as a ratio of mean stem surface area S to the area of an imaginary column having the height H equal to mean tree height and radius R of mean area occupied by a tree. Thus biomass density in terms of stem surface area was given by Inoue as

$$B = S/2\pi RH. \tag{2.9}$$

Then Inoue [10] supposed that mean height H and biomass density B may be related to mean stem surface area S through allometric power relationships

$$H \propto S^\alpha \quad \text{and} \quad B \propto S^\beta. \tag{2.10}$$

At the same time, simple geometrical considerations suggest that the radius R should be related to stand density ρ as

$$R \propto \rho^{-1/2}. \tag{2.11}$$

As a result, from (2.9)–(2.11) Inoue [10] derived his model in the form

$$S \propto \rho^{-1/[2(1-(\alpha+\beta))]}, \tag{2.12}$$

which gives one the opportunity to estimate the slope of the mean stem surface area–stand density relation provided parameters α and β are known.

In order to get numerical values of parameters α and β Inoue [10] used measurement data obtained from even-aged pure stands of Japanese cedar and Japanese cypress. Due to thinning, the stands were differently stocked, which gave the author an opportunity to relate stand density against mean and total stem surface area in these stands. The stem surface area was estimated through relationships received in previous studies (see page 20).

Because self-thinning laws are considered to be an upper boundary limiting mean tree sizes at a given stand density it was necessary to select those forest stands showing this possible maximum, the so-called overcrowded stands. Inoue [10] admitted that there was no objective way to select overcrowded stands among all others, thus he used a visual selection of stands that lay close to the observed upper boundary.

Fitting the log-transformed data against equation $\ln Y = a \ln X + b$ (Y and X being dependent and independent variables, respectively, a and b being parameters) Inoue [10] received the estimations for α to be 0.477 ($r = 0.987$) for cedar and 0.424 ($r = 0.951$) for cypress, with a significant difference found between the species. The estimations for β were 0.064 ($r = 0.451$) for cedar and 0.062 ($r = 0.427$) for cypress, with no significant difference between the species.

Discussing the estimations obtained, Inoue [10] considered an isometric model assuming an isometric tree growth; according to the model mean stem surface area, S should be proportional to the reciprocal of stand density ρ as

$$S \propto 1/\rho, \tag{2.13}$$

and therefore the slope should be -1. As Inoue commented, in the allometric model (2.12), the slope can be -1 only if $\alpha + \beta = 1/2$.

Another consequence of the isometric model is that the total surface area T should be a constant, because $T = S\rho \propto \rho \cdot 1/\rho = 1$. Admitting that the isometric model is a specific case of the allometric model Inoue [10] concluded that the sum $\alpha + \beta$ obtained from the fitting is very close to 1/2 and therefore the total surface area has to be constant for the data. The data analysis showed that the total surface area fell very slightly with the stand density increase for cedar and stayed practically constant for cypress. Inoue suggested that the maximal total surface area amounted to 1.483 ha/ha for cedar and 0.949 ha/ha for cypress overcrowded stands.

One can note that the hypothesis of a maximal total stem surface area for overcrowded forest stands formulated on the basis of static data contains a dynamic sense as well. In fact, data on total stem surface area presented by Inoue [10] and Inoue and Nishizono [11] for differently stocked stands suggest a sort of independence of the total stem surface area of stand density. If such independence *persists in time* and forest stands are subjected to self-thinning (i.e., stand density decreases) then the forest stands should move against stand density in the space "total stem surface area–stand density" (Fig. 2.3).

Fig. 2.3 A fictitious example
of stem surface area for
overcrowded forest stands.
Dashed line denotes a
statistical maximal total stem
surface area, *dots* are
individual stands. *Arrow*
shows time dynamics

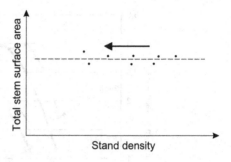

An actual time development of total stem surface area can only be observed
from long-term direct experiments. A remarkable contribution to long-term observation of forest growth was done by levels-of-growing-stock cooperative studies
in Douglas fir(*Pseudotsuga menziesii*) performed by the US Forest Service and
openly published in its working reports [4, 5, 12, 15]. Control sample plots in
the experiments were left to grow without any intervention and were remeasured
repeatedly every few years for several decades. Douglas fir stands on the control
plots are therefore a convenient object to follow development of various measures
of stand size, total stem surface area included. The fragments of Douglas fir data
used below are given in Tables A.1, A.2, A.3, A.4, and A.5 in the Appendix.

Figures 2.4 and 2.5 depict development of total stem surface area on the control
plots in five levels-of-growing-stock experiments: Marshall and Curtis [15], King et
al. [12], and Curtis et al. [4, 5]. The total stem surface area T was estimated by the
author from the published data using the approach described by Inoue [8], that is,
as a product of mean dbh D and mean height of tallest trees H as $T = 0.5\pi DHN$
where N is the stand density. The formula obviously presents a cone approximation.

Initial ages at which the observations started were in a limited range from 19 to
29 years (Fig. 2.4). Meanwhile, the span of initial stand densities was substantially
wide, from 1467 trees/ha (Skykomish) to 4244 trees/ha (Hoskins) (Fig. 2.5). The
three most densely stocked Douglas fir stands (Hoskins, Rocky Brook, and Iron
Creek) show how total stem surface area evolved from increasing to a leveling off
at maximal values. Despite definite initial stocking differences when the stocking
reaches an approximate interval of 2000–2500 trees/ha the values of total stem
surface area are very close for these stands.

Also, the development of total stem surface area not only levels off but shows
a definite decrease after reaching maximum. This gives an opportunity to see three
distinctive stages in the development: growth, more or less prolonged maximum,
and decrease. Because the stages are well differentiated from one another it is
possible to compare the development of total stem surface area with development
of other important measures including volume stock.

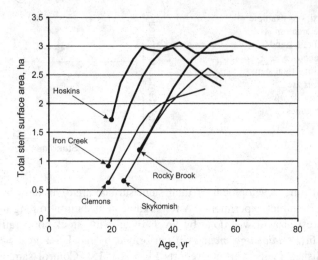

Fig. 2.4 Development of total stem surface area against stand age on control plots in levels-of-growing-stock cooperative studies in Douglas fir [4, 5, 12, 15]. The graphs were constructed by the author from the published tables, with the total surface area being estimated as a product of mean dbh D and mean height of tallest trees H as $T = 0.5\pi DHN$ where N is the stand density. *Filled circles* denote the beginning of dynamics. The name designators are names of the experiments

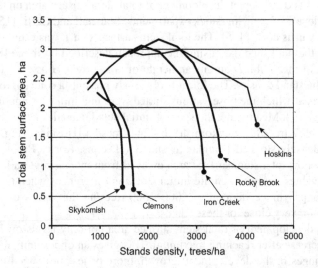

Fig. 2.5 Development of total stem surface area against stand density on control plots in levels-of-growing-stock cooperative studies in Douglas fir [4, 5, 12, 15]. The legends are as in Fig. 2.4

Let us first consider what one can expect from such a comparison theoretically. Suppose that total stem surface area may be expressed through D, H, and N as

$$\hat{S} = \pi \frac{D}{2} \cdot H \cdot N = K_1 \cdot N^\alpha, \tag{2.14}$$

where K_1 and α are constants.

In Eq. (2.14), it is supposed that dependence of the total stem surface area on stand density may be approximated by a power function (right side of the equation). Being a monotonic function the power function is not expected to cover the entire range of values because the real development of total stem surface area has an obvious nonmonotone form. Still for monotonic segments such an approximation may be reasonable (see also page 53).

Expression (2.14) for this particular geometrical model may be easily converted (by multiplying by $\frac{1}{3}D/2$) to the expression for total volume stock $\hat{V}$ as

$$\hat{V} = \pi \frac{1}{3} \frac{D^2}{4} \cdot H \cdot N = K_1 \cdot \frac{1}{3} \frac{D}{2} \cdot N^\alpha. \tag{2.15}$$

Next, according to a geometric model of a forest stand (see Sect. 3.1), stem diameter D may be expressed through N as

$$D = K_2 \left(N^{-\frac{1}{2}\gamma} \right),$$

showing that mean stem diameter is proportional to the square root of the reciprocal of stem density (the more dense the stand the smaller is the diameter), K_2 being a constant.

Expression (2.15) is therefore converted to

$$\hat{V} = K_3 \cdot N^{\alpha - \frac{\gamma}{2}} \tag{2.16}$$

where $K_3 = K_1 K_2 / 6$. Formula (2.16) gives the development of total volume stock at different stages of growth where parameter α stays constant, that is, for monotone segments. For these different monotonic segments, parameter α will naturally be different and reflect the rate and direction of both total stem surface area and total volume stock development with stand density change.

The dependence of the total volume stock increment on α may be seen from the N-derivative of $\hat{V}$:

$$\hat{V}' = K_3 \cdot \left(\alpha - \frac{\gamma}{2} \right) \cdot N^{\alpha - \frac{\gamma}{2} - 1}. \tag{2.17}$$

Because the development of $\hat{S}$ and $\hat{V}$ is dependent on stand density which decreases with time, the meaning of derivatives is reversed. That is, negative α means that total stem surface area increases with decrease of N.

Thus if $\alpha < 0$ (total stem surface area grows with the decrease of density) then $\hat{V}' < 0$ (see (2.17)); that is, total volume stock grows as well, which is a natural expectation. If $\alpha = 0$ (total stem surface area stays constant) then $\hat{V}' < 0$ (total volume stock continues to grow). Lastly, if $\alpha > 0$ (total stem surface area decreases) then $\hat{V}' < 0$ or $\hat{V}' > 0$ is dependent on values of α and γ; that is, total volume stock may increase or decrease.

To interpret these considerations, calculations suggest that when total stem surface area stops growing ($\alpha = 0$) the total volume stock still continues to increase ($\hat{V}' < 0$). Only when the total stem surface area decreases ($\alpha > 0$) can the total volume stock also decrease.

These theoretical inferences may be verified against the field data of the Hoskins experiment [15] because this experiment shows distinctive stages of total stem surface area development: growth, constancy, and decrease (Fig. 2.5).

Figure 2.6 depicts the development of total volume stock and total stem surface area against stand density in the Hoskins experiment. As can be seen from the figure, graphs support the theoretical considerations. The relative development of the measures in question is so that the growth leveling off and subsequent decrease are seen *first* in total stem surface area and only *later* in total volume stock. This means that the total stem surface area may serve as an indicator of total volume stock changes to come in the near perspective, which may be used as a prognosis tool in forest stand management.

To finish the chapter, it is relevant to return briefly to the matter of increment because stand increment was one of the stimuli to measure and study stem surface area. As is well known, forest science differentiates between two measures of increment, current increment and mean increment. Current increment Z_c is given

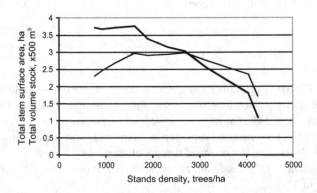

Fig. 2.6 A comparison of development of total volume stock (*thick line*) and total stem surface area (*thin line*) for the Hoskins experiment [15]. The values of total volume stock are taken from the published tables and scaled by dividing by 500 for convenience of representation. Values of total stem surface area are calculated by the author as given in Eq. (2.14)

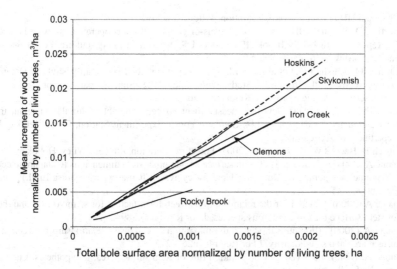

Fig. 2.7 Relation between mean increment in Douglas fir stands [4, 5, 12, 15] and stem surface area. Both variables are normalized by the number of living trees. Stem surface area is calculated through quadratic mean diameter (see explanations in the text)

as $Z_c = (V_{i+1} - V_i)/(A_{i+1} - A_i)$, where V is volume stock, A is age, and i denotes a moment of time. Mean increment Z_m is simply volume stock V divided by age A as $Z_m = V_i/A_i$.

Both increments were plotted against total stem surface area for the Douglas fir data. No good relationship was found between current increment and total stem surface area because this kind of increment appeared to be quite a changeable variable. The mean increment, however, showed a rather promising relation to total stem surface area as given in Fig. 2.7. For the sake of compatibility with the volume stock variable, the calculation of total stem surface area was done using the quadratic mean diameter as $\hat{S} = 0.5\pi D_q \cdot H \cdot N$, where D_q is the quadratic mean diameter.

As shown in Fig. 2.7, the relation of mean increment to total stem surface area, both normalized by the current stand density, is very close to a simple linear function. The slopes appear to be individual for different forest stands, which may be a reflection of local growth conditions. Thus it may be possible to estimate mean increment in a forest stand if one has a value of normalized stem surface area and the slope of the relation for a particular forest stand.

References

1. Anuchin NP (1982) Forest mensuration: textbook for high school. Lesnaya promyshlennost, Moscow (in Russian)
2. Carron LT (1968) An outline of forest mensuration: with special reference to Australia. Australian National University Press, Canberra

3. Chapman HH (1921) Forest mensuration. Wiley, New York
4. Curtis RO, Marshall DD et al (2009) Levels-of-growing-stock cooperative study in Douglas-fir: Report No. 18-Rocky Brook, 1963–2006. US Department of Agriculture, Forest Service, Pacific Northwest Research Station
5. Curtis RO, Marshall DD et al (2009) Levels-of-growing-stock cooperative study in Douglas-fir: Report No. 19-the Iron Creek Study, 1966–2006. United States Department of Agriculture, Forest Service, Pacific Northwest Research Station
6. Grosenbaugh LR (1954) New tree-measurement concepts: height accumulation, giant tree, taper and shape. Occas, paper 134. Southern Forest Experiment Station, Forest Service, US Department of Agriculture
7. Husch B, Beers TW, Kershaw JAJ (2003) Forest mensuration, 4th edn. Wiley, Hoboken
8. Inoue A (2004) Relationships of stem surface area to other stem dimensions for Japanese cedar (Cryptomeria Japonica D. Don) and Japanese cypress (Chamaecyparis obtusa Endl.) trees. J For Res 9(1):45–50
9. Inoue A (2006) A model for the relationship between form-factors for stem volume and those for stem surface area in coniferous species. J For Res 11(4):289–294
10. Inoue A (2009) Allometric model of the maximum size–density relationship between stem surface area and stand density. J For Res 14(5):268–275
11. Inoue A, Nishizono T (2015) Conservation rule of stem surface area: a hypothesis. Eur J For Res 134(4):599–608
12. King JE, Marshall DD, Bell JF (2002) Levels-of-growing-stock cooperative study in Douglas-fir: Report No. 17-the Skykomish study, 1961–1993; the Clemons study, 1963–1994. Pacific Northwest Research Station, USDA Forest Service
13. Kira T, Shidei T (1967) Primary production and turnover of organic matter in different forest ecosystems of the western Pacific. Jpn J Ecol 17(2):70–87
14. Lexen B (1943) Bole area as an expression of growing stock. J For 41(12):883–885
15. Marshall DD, Curtis RO (2001) Levels-of-growing-stock cooperative study in Douglas-fir: report no. 15-Hoskins: 1963–1998. United States Department of Agriculture, Forest Service
16. Pokornỳ R, Tomášková I (2007) Allometric relationships for surface area and dry mass of young Norway spruce aboveground organs. J For Sci 53(12):548–554
17. Prodan M (1965) Holzmesslehre. JD Sauerländer, Frankfurt am Main
18. Swank WT, Schreuder HT (1974) Comparison of three methods of estimating surface area and biomass for a forest of young eastern white pine. For Sci 20(1):91–100
19. Van Laar A, Ak̦ca A (2007) Forest mensuration, vol 13. Springer Science & Business Media, Dordrecht

Chapter 3
Self-thinning and Stem Surface Area

Abstract Any object of study is a number of variables, or measures, whose values are delivered by corresponding direct or indirect measurements. These variables are often dependent on time and other variables, which produces relationships of variables to time and to each other.

In this chapter, modeling of alterations in the stem surface area is used to study two questions. The first is about whether "secondary" relationships may exist in the structure of a forest stand as a system of growing and competing trees. Definitions of "primary" and "secondary" relationships are given in the section, Primary and Secondary Relationships: Look Through a Geometrical Model of Forest Stand, below. Here a model is required that predicts an interplay of relationships of different variables in a forest stand. The model should be transparent enough to allow an analytical consideration. On the other hand, it should bear enough similarity with real forest stands to allow a comparison with field data.

The second question deals with understanding the current status of the famous "$-3/2$ rule." It turns out that a model based on stem surface area helps to clarify the place of the rule in the theory of self-thinning and further to illustrate similarities in the self-thinning within tree species.

Keywords Cone approximation • Douglas fir • Geometrical model • Primary relationships • Scots pine • Secondary relationships • Stem surface area

3.1 Primary and Secondary Relationships: Look Through a Geometrical Model of Forest Stand

For many years it has been observed that growth within a dense forest stand alters the stature of trees compared with trees growing in the open. It is these alterations that are the most vivid evidence of interactions in the stand and that bring about relationships between variables describing the stand.

Let uppercase letters such as A, B, C, D be some variables. Then functions such as F_1, F_2, F_3 may give an idea of relationships among the variables, that is, of how the variables relate to each other, for example, $A = F_1(B), B = F_2(C), C = F_3(D)$. The functions applied in research are dependent as a rule on constants, or parameters,

© The Author(s) 2017
V.L. Gavrikov, *Stem Surface Area in Modeling of Forest Stands*,
SpringerBriefs in Plant Science, DOI 10.1007/978-3-319-52449-8_3

that are important parts of the mathematical constructions. Equations describing relationships between the variables can thus be written as

$$A = F_1(B, a_1, a_2, \ldots), B = F_2(C, b_1, b_2, \ldots), C = F_3(D, c_1, c_2, \ldots), \qquad (3.1)$$

where the lowercase letters a, b, c stand for parameters of the corresponding functions.

If A, B, C, D are simple quantitative measures of a forest stand, such as stem density per unit area, mean diameter, mean height, and the like, relationships (3.1) may be termed "primary" because they relate one immediately observable variable to another.

On the other hand, if one considers the forest stand to be a system he or she may hypothesize that another sort of relationship may also exist, when one relationship is linked to the other relationship. To give a specific example, parameters of one primary relationship may be linked to parameters of another primary relationship as

$$a_1 = f_1(b_1), b_2 = f_2(c_2), \text{etc.,} \qquad (3.2)$$

where f_1, f_2 are functions that interrelate the parameters of different relationships. Such types of "relationship of relationships" given in (3.2) may be termed "secondary" relationships.

Is a question of the possibility of secondary relationships a result of idle curiosity? Is it not sufficient to know that primary relationships can be defined and explored?

The advantage of knowing the secondary relationships has the same source as in the case of primary relationships. It is the ability to predict. Primary relationships give us the ability to predict, with a certain degree of accuracy, one variable from another. Secondary relationships, if they do exist, allow us to predict a relationship form from known parameters of another relationship.

In what follows, a simple geometrical model of a stand is first defined and its properties are derived. Second, the model is corrected to make it more realistic and to derive how the corrections influence the model properties. Third, the model is examined to uncover how well it predicts the dynamics of the stem surface area in real forest data from the book by Usoltsev [38] who gathered from the literature a large collection of data (altogether over 10,000 descriptions) on the phytomass of various fractions in forests, mostly from European and Asian studies. All this will help in answering the question about secondary relationships.

3.1.1 A Geometrical Model of Forest Stand

Let us consider a population of cones (Fig. 3.1) that is defined by the following set of conditions[5].

1. N Cones stand closely spaced and occupy an area S.

2. The cones are narrow and high enough to accept that the generatrix l is approximately equal to cone height h, that is, $l \approx h$. (In fact, this is a matter of pure convenience to express the surface; Fig. 3.1b.)
3. If the quantity of the cones decreases their horizontal size grows accordingly (Fig. 3.1c, d).
4. Thus cone base radius r may be at every moment estimated by N and S (Fig. 3.1a) as

$$r = \frac{1}{2}\sqrt{\frac{S}{N}}. \qquad (3.3)$$

With respect to surface area the model has an obvious property: if $\hat{S}$ is the total surface area (excluding the area of the bases) of the cone population then

$$\hat{S} = \pi \cdot l \cdot r \cdot N. \qquad (3.4)$$

If, next, one imposes a constraint on $\hat{S}$ demanding that it be a constant, $\hat{S} = C =$ const, then one comes to the relation

$$l = \frac{C}{rN\pi} = \frac{rC}{r^2N\pi} = \frac{rC}{\dfrac{1}{4}\dfrac{S}{N}N\pi} = \frac{4C}{\pi S} \cdot r. \qquad (3.5)$$

Relation (3.5) is quite an important point of the analysis because it means that imposing the constraint $\hat{S} = C = $ const on the cone population brings about a relationship $l(r)$. In this particular case, this is the linear relationship of the type $l = k \cdot r$, where k is a constant. And vice versa, imposing the relationship $l = k \cdot r$ on

Fig. 3.1 A geometrical model of forest stand[5]: (**a**) calculation of the radius of a cone base; (**b**) relationship between the height h and generatrix l; (**c**)–(**d**) a decrease in the number of cones leads to growth of their sizes

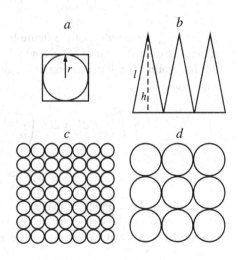

the population immediately brings about the equality $\hat{S} = $ const. Also, relation (3.5) describes the well-known condition of the geometric similarity.

It should be noted that the model does not incorporate time. Instead, the role of time is played by number N of the cones. It is implied that N can only decrease, which resembles a usual dynamics of a dense even-aged forest stand.

In the context, relation (3.5) reflects interdependence of the radius and the generatrix in the process of growth, not a tapering of the radius along the tree axis.

3.1.2 Introducing Realism into the Model

Obviously, the model is far from being an exact reflection of a forest stand. Trees do not stand close to each other. However, the condition (3.3) may be loosened by introducing a factor in order that a tree does not occupy the entire area (Fig. 3.1a).

Two other suppositions are examined in the following upon which the model is based: (1) the very assumption of conical shape of the "trees" and (2) the relationship (3.3) between r and N.

1. Whether trees are cones. Let the generatrix of the figures not be a straight line but rather a nonlinear power function, which is a tradition in studies of bole tapering. Thus,

$$r = r(l) = k \cdot l^{\alpha}, \tag{3.6}$$

where l stands for distance from the tip of the bole, r stands for the radius, and k and α are parameters. The obvious boundary conditions for the case are

$$l_0 = 0; r_0 = kl_0^{\alpha} = 0; r_{max} = kl_{max}^{\alpha}, \tag{3.7}$$

where r_{max} is the radius of the figure base and l_{max} is its total stem length. The circumference at distance l is given by $k \cdot 2\pi r = k \cdot 2\pi l^{\alpha}$.

Integrating the latter expression from l_0 to l_{max} gives the surface of the figure as

$$2k\pi \int_{l_0}^{l_{max}} l^{\alpha} dl = 2k\pi \left(\frac{l_{max}^{\alpha+1}}{\alpha + 1} - \frac{l_0^{\alpha+1}}{\alpha + 1} \right) = \frac{2k\pi}{\alpha + 1} l_{max}^{\alpha+1}$$

$$= \frac{2k\pi}{\alpha + 1} l_{max}^{\alpha} \cdot l_{max} = \frac{2k\pi}{\alpha + 1} l_{max}^{\alpha} \cdot r_{max}.$$

It should be mentioned that a usual assumption for such an integrating procedure would be that $dr/dl = 0$ (for an estimation procedure for the stem surface area as a body of revolution see, e.g., [9]).

The total surface area will then be

$$\hat{S} = \frac{2k\pi}{\alpha + 1} \cdot r_{max} \cdot l_{max} \cdot N \tag{3.8}$$

which differs from (3.4) only by a constant factor. Therefore, the power law nonlinearity of the generatrix will not qualitatively change the relationship (3.5).

Another consideration of the difference of trees from cones comes from instrumental measurements of the stem surface areas in various trees. It has been shown in Chap. 2 that the ratio "surface areas estimated by cone formula"/"genuine surface areas" oscillates in a narrow range of 0.79–0.86 after 20–30 years of age in Scots pine. Inoue [9] reported (see also Chap. 2) a close relationship of the stem surface area s and the product of the total height h and diameter at breast hight d. For Japanese cedar, the relationship looked like $s = 1.937 \cdot dh$ and for Japanese cypress $s = 1.921 \cdot dh$. Converting the relation, for example, for the cedar, into a cone formula gives

$$\pi \cdot s = 1.937 \cdot 2\pi rh; \quad \frac{\pi rh}{s} = \frac{\pi}{1.937 \cdot 2} \approx 0.81,$$

which is rather close to the estimations for Scots pine. Inoue also mentioned that the slopes of the relationships found may have small variation among conifers. All this means that the cone approximation may be used in modeling the surface area implying that the genuine surface area is often different from the cone surface area by a roughly constant factor.

Thus, although trees are surely not cones, a cone approximation in regard to stem surface area does not distort the results of the analysis. Based on the above considerations, cone approximation is used below in further calculations.

2. Whether the relationship (3.3) is realistic. To verify this, it is enough to look at the dynamics of such a measure as the basal area of a stand. The problem of the relationship (3.3) is that it brings about the independence of basal area G from stem density N:

$$G = \pi \cdot r^2 \cdot N = \pi \frac{1}{4} \frac{S}{N} \cdot N = \frac{\pi S}{4} = const,$$

which is clearly not realistic. A yield table would directly show that the basal area of a stand grows, mostly as a convex function, with the reduction in density. It means that an assumption of the geometric similarity is not enough to model a stand dynamics. Empirical observations show that r grows with the decrease in N not as $r \propto \sqrt{1/N}$ but faster. To correct this, an assumption may be considered that the exponent of power at N is not equal to unity in (3.3) so that

$$r = \frac{1}{2}\sqrt{\frac{S}{N^\gamma}}. \tag{3.9}$$

Preliminary estimations show that γ often lies in the range from 1 to 2 for real data.

The relationship (3.9) modifies (3.5) qualitatively because it introduces a nonlinear relation in it. It is easy to see that in the case of (3.9)

$$\hat{S} = \pi \cdot r \cdot l \cdot N = \frac{\pi \cdot r^2 \cdot l \cdot N}{r} = \frac{\pi \cdot l \cdot \frac{1}{4}\frac{S}{N^\gamma} \cdot N}{r} = \frac{\pi \cdot l \cdot S}{4 \cdot N^{\gamma-1}r}. \qquad (3.10)$$

Imposing condition $\hat{S} = C = \text{const}$ on (3.10) and expressing N from $r^2 = \frac{1}{4}\frac{S}{N^\gamma}$ gives

$$l = \frac{r^{\frac{2}{\gamma}-1} \cdot C}{\pi \cdot \left(\dfrac{S}{4}\right)^{1/\gamma}}. \qquad (3.11)$$

If from empirical estimations of $r(N)$ $2 > \gamma > 1$ then $2 > \dfrac{2}{\gamma} - 1 > 0$, which makes the relationship $l \propto r^{\frac{2}{\gamma}-1}$ quite realistic because the dependence of tree height on trunk diameter is often described by a concave function for younger trees and a convex function for older trees in forest science.

It is worth mentioning that there exist primary relationships of $r = F_1(N, a, \ldots)$ (3.9) and $l = F_2(r, b, \ldots)$ (3.11), where F_1 and F_2 are functions and a and b are exponents of the functions. It is easy to see from (3.9) and (3.11) that the exponents are linked to each other through a common term γ and may be predicted from one another for the case $\hat{S} = C = \text{const}$. It means that a secondary relationship of the sort $a = f(b)$ may exist for this particular case.

In other words, one could imagine a logical triangle in which any two known assertions unambiguously define the third one. Formally, the model analysis may be summarized as follows. There are three relationships:

$$\hat{S} = C = \text{const}, \qquad (3.12)$$

$$r \propto \sqrt{\frac{1}{N^{\gamma_1}}} \quad \text{or} \quad \left(N^{-\frac{1}{2}\gamma_1}\right), \qquad (3.13)$$

$$l \propto r^{\frac{2}{\gamma_2}-1}. \qquad (3.14)$$

where inferior indices at γ_1 and γ_2 mean that the same parameter γ is estimated from two different relationships. The model itself therefore has such a property that the relations (3.12)–(3.14) are closely connected to each other. Each two determines the third one. For example, if (3.12) is true and the power exponent in (3.14) is known then one can predict the power exponent in (3.13) because in this case $\gamma_1 = \gamma_2$. If $\gamma_1 \neq \gamma_2$ in (3.13) and (3.14), respectively, then one can expect the condition (3.12) to be not valid.

3.1.3 Comparing the Model Against Forest Data

As has been said above, the model predictions were compared to forest field data collected by Usoltsev [38]. Before introducing the results of the comparison, a few preliminary notes have to be given. First, the data in the database are presented as is traditionally done in forest science: in the form of mean values of various variables. As usual, only the average diameter at breast height D, mean height H, and stand density N are given for every sample plot. Theoretically, having the data at one's disposal it is possible to estimate the total stem surface area BS by

$$BS = \frac{1}{2}\pi \cdot D \cdot H \cdot N. \qquad (3.15)$$

Earlier, Inoue [9] found the relationship (3.15), multiplied by a species-specific factor, to be a useful approximation to the stand surface area.

However, from statistics it is known that for two variables X and Y with corresponding means $\overline{X}$ and $\overline{Y}$ the relation $\overline{X \cdot Y} = \overline{X} \cdot \overline{Y}$ is only exact if X and Y are completely independent of each other. That is why it would be preferable not to call (3.15) a total stem surface area of a forest stand but an estimation of surface area (ESA) that approximates the genuine surface area and reflects it.

Second, the data do not completely match the model because they represent a static collection of sample plots whereas the model implies the dynamics of a single stand. Therefore, the more the data resemble the dynamics of a single stand the more they are suitable for comparison.

For the sake of comparability, the database by Usoltsev [38] was searched only for Scots pine data because *Pinus sylvestris* L. is one of the most commonly distributed tree species in boreal Eurasia. In the case of natural stands they had to be pure pine stands. Also, only the data for site index I (the highest forest productivity in the Russian bonitation practice) were taken.

Because of the static nature of the Usoltsev [38] data it would be worthwhile also to compare the model to dynamic data that are the result of long-term measurements of the same forest stand. Such kinds of data are not so abundant in the forest science literature but some of them are still available for analysis.

In the time span from the 1960s to the beginning of the twenty-first century, a number of experiments have been performed by the US Forest Service that received the name of "levels-of-growing-stock cooperative study in Douglas fir". The experiments were aimed at studying forest stand growth reactions to various thinning practices. Every experiment had a certain number of control plots in which no thinning interventions were applied. For the comparison with the model, the most prolonged Hoskins experiment was taken that lasted from 1963 to 1998. The report [18] of the experiment contains various data on stand development including mean diameters at breast height, mean height of 100 tallest trees, stem density, and other stand measures (see also Tables A.1, A.2, A.3, A.4, and A.5 in the Appendix). For the data, total stem surface area was calculated using (3.15) as well.

The approximations for the relations (3.13) and (3.14) were performed with the help of STATISTICA 6 software.

3.1.3.1   Dataset #1

The dataset is marked in Usoltsev [38, p. 239] as published by Mironenko in 1998 (below, the data are referred to as Mironenko-98; see Appendix). These are pine plantations from the forest-steppe zone (Tambov administrative region, Russia); the range of ages is from 70 to 150. The graphical representations of the data are shown in Fig. 3.2.

From the data presented, one can see that the estimation of total surface area stays roughly constant with the reduction in the stand density (Fig. 3.2a). The model analysis as reflected in (3.12)–(3.14) says that one can predict γ_1 from γ_2. In fact, $\dfrac{2}{\gamma_2} - 1 = 0.6402$ (Fig. 3.2b); that is, $\gamma_2 = \dfrac{2}{1.6402} = 1.2194$, standard error (SE) = 0.0225.

On the other hand, $-\dfrac{1}{2}\gamma_1 = -0.6144$ (Fig. 3.2c); that is, $\gamma_1 = 2 \cdot 0.6144 = 1.2288$, SE = 0.0068. Therefore it may be accepted with a certain degree of accuracy (about 0.75%) that γ_1 roughly equals γ_2 for this dataset, which conforms to the model analysis (3.12)–(3.14).

3.1.3.2   Dataset #2

The dataset is marked in Usoltsev [38, p. 31] as published by Kozhevnikov in 1984 (below, the data are referred to as Kozhevnikov-84, see Appendix). These are pine plantations in Belorussia; the range of ages subsampled for the calculations is from 15 to 60. The graphical representations of the data are shown in Fig. 3.3.

By contrast with dataset #1, one could not accept a supposition that total stem surface area was constant with variations of the stand density. Rather, the tendency rises with the fall of the density (Fig. 3.3a). Therefore, based on the theory, the estimations of γ_1 and γ_2 should visibly differ from one another. Their estimations are: $\dfrac{2}{\gamma_2} - 1 = 0.8464$ (Fig. 3.3b); that is, $\gamma_2 = 1.0832$ (SE = 0.0233) and $-\dfrac{1}{2}\gamma_1 =$

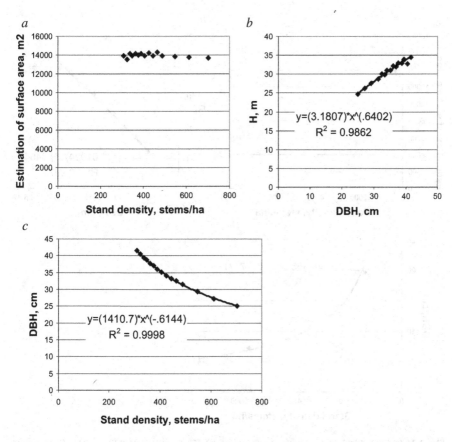

Fig. 3.2 The graphical representations of dataset #1 Mironenko-98 [38, p. 239]. The graphs were constructed by the author on the basis of the published tabular data. Each point represents a sample plot. (**a**) Development of surface area estimation with the drop of stand density, (**b**) relation between DBH and mean height H, (**c**) growth of DBH with the drop of stand density

-0.717 (Fig. 3.3c); that is, $\gamma_1 = 1.434$ (SE = 0.092), which means a difference of about 24%.

3.1.3.3 Datasets #3–#10

The estimations of γ_1 and γ_2 for other Scots pine data are summarized in Table 3.1. For a correct treatment of the data, there should be a clear tendency in the relation $\hat{S}(N)$. As noted above, the data are in fact static measurements gathered, perhaps occasionally, from many forest stands. In some datasets therefore a subsampling had to be done to extract the tendency in the relation between the estimation of surface area and the stand density (SD).

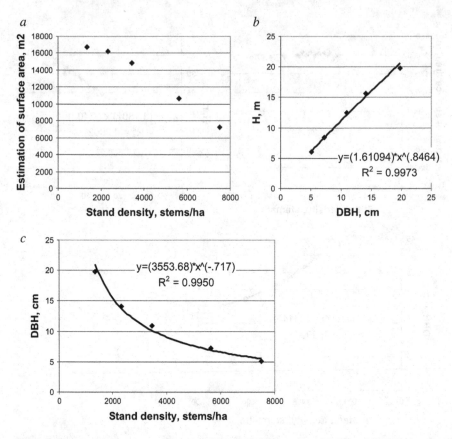

Fig. 3.3 The graphical representations of dataset #2 Kozhevnikov-84 [38, p. 31]. Each point represents a sample plot. (**a**) Development of surface area estimation with the drop of stand density, (**b**) relation between DBH and mean height H, (**c**) growth of DBH with the drop of stand density

An overview of Table 3.1 shows that the relation $\gamma_1 > \gamma_2$ correlates with a rising tendency and $\gamma_1 < \gamma_2$ correlates with a falling tendency. This means that in the case of a flat tendency a relation $\gamma_1 \approx \gamma_2$ would be a plausible one. In fact, datasets #1, 9, and 10 give examples that γ_1 and γ_2 are much closer to each other than in other cases.

The results presented in Table 3.1 can also be shown visually as in Fig. 3.4. In this figure, each dataset is located on the plane γ_1 against γ_2 and the straight diagonal line denotes the places where $\gamma_1 = \gamma_2$. The question is to confront the dataset positions on the plane with the line. It can be seen that all the datasets with a rising $\hat{S}(N)$ tendency are lying to the right of the line, where $\gamma_1 > \gamma_2$. On the opposite side, all the datasets with falling tendency are lying to the left of the line, that is, where $\gamma_1 < \gamma_2$. The datasets with a flat tendency, which means that $\hat{S}$ is approximately constant with the reduction of N, are lying very close to the line showing that $\gamma_1 \approx \gamma_2$ in those cases.

Table 3.1 A summary of the datasets used to illustrate the model

Dataset	Reference in Usoltsev [38]	Tendency of ESA with SD, slope[a]	$\gamma_1 \pm SE$[b]	$\gamma_2 \pm SE$
1.	Mironenko-98, Russia, Tambov administrative region, p. 239, range of ages 70–150	Flat, −0.6	1.2288 ± 0.0068	1.2194 ± 0.0225
2.	Kozhevnikov-84, Belorussia, p. 31, range of ages (subsample) 15–60	Rising, −1.6104	1.434 ± 0.092	1.0832 ± 0.0233
3.	Gruk-79, Belorussia, p. 30, range of ages 10–40	Rising, −2.3842	1.522 ± 0.1424	0.8818 ± 0.0343
4.	Uspensky-87, Russia, Tambov, p. 240, range of ages (subsample) 10–30	Rising, −2.3671	1.8092 ± 0.22	0.8981 ± 0.02661
5.	Gabeev-90, Russia, Novosibirsk, p. 482, range of ages 10–50	Rising, −3.4437	1.8464 ± 0.1424	0.9727 ± 0.0416
6.	Heinsdorf and Krauß-90, p. 56, Germany, Eberswalde, range of ages (subsample) 50–120	Falling, 3.4428	1.0168 ± 0.0133	1.2769 ± 0.0076
7.	Uspensky-87, Russia, Tambov, p. 240, range of ages (subsample) 80–120	Falling, 3.4274	1.038 ± 0.0072	1.1811 ± 0.0034
8.	Kurbanov-02, Russia, Yoshkar-Ola, p. 211, range of ages 68–128	Falling, 6.9756	0.8718 ± 0.108	1.2887 ± 0.0908
9.	Uspensky-87, Russia, Tambov, p. 240, range of ages (subsample) 30–60	Flat, −0.0794	1.1026 ± 0.0072	1.1041 ± 0.009
10.	Lebkov and Kaplina-97, Russia, Vladimir, p. 203, range of ages (subsample) 25–77	Flat, −0.468	1.1328 ± 0.0704	1.1209 ± 0.0364

For initial data see Appendix Tables A.6, A.7, A.8, A.9, A.10, A.11, A.12, and A.13
[a]Slope is seen as variation of ESA with the decrease of stand density. That is why the rising tendency has a negative slope and the falling tendency has a positive slope
[b]Standard error

Furthermore, a rising tendency should be associated with younger forest stands because they usually have higher growth potential allowing them to compensate for loss of stems during self-thinning. In much the same way the falling tendency should associate with older forest stands. In fact, mostly relatively younger forest stands, aged under 60 years, are represented on the right side in Fig. 3.4 whereas on the left side there are somewhat older forest stands of ages up to 120 and more years. It would therefore be logical to hypothesize that even-aged forest stands follow the path from $\gamma_1 > \gamma_2$ through $\gamma_1 \approx \gamma_2$ to $\gamma_1 < \gamma_2$ in their natural course of growth and self-thinning.

Fig. 3.4 Representation of the results of Table 3.1 on a plane γ_1 against γ_2. The *straight diagonal line* denotes the places where $\gamma_1 = \gamma_2$. Each dataset is represented by a point and the numbers on the graph are the numbers of the datasets

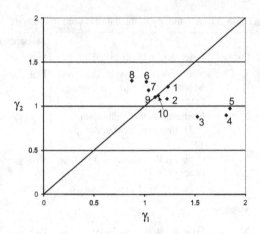

3.1.3.4 Douglas Fir Dataset

This hypothesis may be tested with the help of the mentioned data on the natural growth of Douglas fir forest stands. In the dataset published by Marshall and Curtis [18], a total of ten measurements covering ages from 20 to 55 years is available. The stem density decreased from 4244 to 757 stems per ha during the time span. In the course of stem density decrease, three clear stages of estimated total surface area *ESA* may be recognized, a rising tendency as in Fig. 3.5(1a); a flat tendency, Fig. 3.5(1b); and a falling tendency, Fig. 3.5(1c). The parameters γ_1 and γ_2 were statistically estimated and the results are shown in Fig. 3.5(2). It is obvious from the figure that the relation between γ_1 and γ_2 develops from $\gamma_1 > \gamma_2$ through $\gamma_1 \approx \gamma_2$ to $\gamma_1 < \gamma_2$ in the course of life of the same forest stand.

The above results show that the stage when $\gamma_1 \approx \gamma_2$ is an inevitable one in the course of self-thinning of an even-aged forest stand. The stage coincides with the condition of the total stem surface area being approximately independent of the change in stem density. The field data therefore support the results (3.12)–(3.14) obtained in analysis of the model, which means that secondary relationships may exist, the relationships between parameters of different primary relationships, in this particular case, of the primary relationships $r(N)$ and $l(r)$.

Apart from the empirical evidence, it would also be useful to outline a theoretical approach to generalize the relationships between γ_1 and γ_2 for any possible dependence of $\hat{S}$ on N. Suppose the development of the total stem surface area can be approximated by a power function, that is,

$$\hat{S} = \pi \cdot r \cdot l \cdot N = K \cdot N^{\alpha}, \tag{3.16}$$

where K and α are parameters. It should be stressed that the power function (3.16) is assumed to approximate only fragments with a uniform tendency (rising, flat, or falling). Clearly, $\alpha < 0$ gives a rising tendency, $\alpha = 0$ a flat, and $\alpha > 0$ a falling tendency.

Fig. 3.5 Results of analysis of Douglas fir data [18]. (1) Dynamics of estimated total surface area with the decrease in stem density. The *straight lines* denote tendencies of *ESA(SD)* development: *a*—rising, *b*—flat, *c*—falling tendency. (2) Development of the Douglas fir stand on the plane γ_1 against γ_2: *a*, *b*, and *c* as in (1)

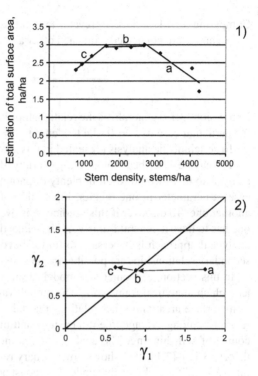

Furthermore, from (3.16) one gets an expression for $l(r)$:

$$l = \frac{K}{\pi \cdot r \cdot N^{1-\alpha}}.$$

(3.17)

Because the value of $\dfrac{1}{N^{1-\alpha}}$ can be expressed from (3.9) as $\dfrac{1}{N^{1-\alpha}} = \dfrac{r^{\frac{2}{\gamma_1}(1-\alpha)}}{\left(\dfrac{S}{4}\right)^{\frac{1-\alpha}{\gamma_1}}}$

the relation (3.17) transforms to

$$l = \frac{K}{\pi \cdot \left(\dfrac{S}{4}\right)^{\frac{1-\alpha}{\gamma_1}}} r^{\frac{2}{\gamma_1}(1-\alpha)-1}.$$

(3.18)

Comparing (3.14) and (3.18) reveals that in the case of $\hat{S} = K \cdot N^\alpha$ the parameters γ_1, γ_2, and α are connected with each other as

$$\frac{1 - \alpha}{\gamma_1} = \frac{1}{\gamma_2} \Rightarrow \gamma_2 = \frac{\gamma_1}{1 - \alpha}. \qquad (3.19)$$

Expression (3.19) generalizes the relationships between γ_1 and γ_2. It is clear that for the particular case of $\alpha = 0$ (flat tendency) $\gamma_1 = \gamma_2$.

To conclude the analysis presented in this section, a forest stand or a set of forest stands may be described though a large variety of relationships. Applying regression analysis to them would result in plenty of parameter estimations. However, in most cases a parameter in one relationship could not be used to predict a parameter in another one. To discover if this prediction is ever possible and to be able to do that one has to have a general framework that links the relationships. In other words, an analytical approach is necessary to find if beyond the primary relationships some secondary relationships are possible in the structure of a forest stand as a system.

In this section, a geometrical model of an even-aged/sized forest stand served as such an analytical framework. The model was able to link variations of the total stem surface area with other relations that take place in the stand, particularly the relations among average tree dimensions and the reduction in stand density in the course of self-thinning. The analysis of the model made it possible to formulate theory (3.12)–(3.14) that shows how primary relationships $\hat{S}(N)$, $r(N)$, and $l(r)$ are interrelated. It was shown that under $\hat{S} = $ const power exponents in $r(N)$ and $l(r)$ can be translated from one to another through a common term γ. It means that relations $\hat{S}(N)$, $r(N)$, and $l(r)$ are a sort of triangle where, knowing any two of the relations, one can predict the third one.

The theory was developed using a basic assumption that the total stem surface area may be constant while the forest stand undergoes self-thinning. It might seem then that the theory applies only to such extreme cases. The assumption, however, is used as a mere reference point to compare with other cases of more frequent occurrence when total stem surface area develops in a way with the fall of forest stand density. The model analyzed allows a kind of generalization that provides a connection of γ_1 with γ_2 also for the cases of nonconstancy of the total stem surface area.

Inoue [10] developed an allometric model of the maximum size density that related stem surface area to stand density. To derive the model, an allometric relationship between mean tree height H and mean surface area S, that is, $H \propto S^\alpha$, on the one hand, and a relationship between biomass density B and mean surface area S, that is, $B \propto S^\beta$, have been considered. It was found that when $\alpha + \beta \approx 1/2$ total stem surface area becomes constant, independent of stand density. In other words, in the case of constant total stem surface area the allometric exponents can be predicted from one another, which is in accord with findings presented in this section.

In spite of the very basic assumptions, the geometrical model developed proved to be workable in many cases of real forest data. The data cases are rather rare where estimations of surface area (*ESA*) are independent of stand density (*SD*). Still, some of them can be found and in those cases the power exponents in relations *DBH(SD)* and *H(DBH)* can be predicted from one another with good accuracy. If *ESA* is clearly dependent on *SD* then the elements γ_1 in (3.13) and γ_2 in (3.14) of the power exponents are sufficiently different. Moreover, in all cases when *ESA* grows with the decrease in *SD* (rising tendency in Table 3.1) the term γ_1 is always larger than γ_2. And vice versa, when *ESA* decreases with the decrease of *SD* (falling tendency in Table 3.1) the term γ_1 is always smaller than γ_2.

It might appear that the choice of examples in Table 3.1 is a bit selective. That is, for younger ages a rising tendency always occurs and for older ages a falling tendency always occurs. Thus it might seem that the relationship between γ_1 and γ_2 shown here depends only on age. In fact, however, the age when *ESA* is roughly constant does not have to be the same (cf. datasets #9 and #10 with dataset #1). Another consideration is that the theory deals with tree sizes and their geometry which in many cases reflect the competitive status of trees better than age. At the same time, the static data from Table 3.1 do not contradict the dynamic data of development of the same forest stand with time (age) as shown in the example of the Douglas fir data (Fig. 3.5).

The relationship between γ_1 and γ_2 may be interpreted from a biological viewpoint. At every moment of time, forest dynamics is competition of the growth and the dieback. Higher values of γ_1 are associated with a weaker diameter growth as the density decreases (see (3.13)). But lower values of γ_2 are associated with a stronger height growth as the diameter increases (see (3.14)). Therefore, the relation $\gamma_1 > \gamma_2$ may reflect that fact that, in terms of the stem surface area, the height growth can set off a loss in density and provide the increase in total stem surface area.

That γ_1 and γ_2 are closely interdependent at $\hat{S} = $ const reflects the time of the highest degree of competition. At that time interval, the total stem surface area is not only constant with density variations but, more important, it is the maximal possible, which may provide evidence of the limitations imposed by resource scarcity. The trees can still grow quickly within the interval so that a loss in density could be compensated by growth, with the compensation being in strict accord with the amount of density reduction.

A possible limitation of the theory developed may be that it touched only the power exponents of the relations γ_1 and γ_2. Other coefficients have not yet been studied. However, even such a simplistic analysis provides some insight into what makes all forest stands similar and therefore can be regarded as a kind of similarity theory in forest studies. Another limitation is that the theory in its current form is applicable mostly to even-sized forest stands, that is, even-aged natural stands or plantations. Still, it shows for at least such forest stands a possibility of subtle relations among the variables and their parameters, the secondary relationships between power exponents of different primary relationships, which enhances the understanding of internal structure and competition in forest stands.

3.2 Status of $-3/2$ Rule and Similarity in Self-thinning

Looked at from a broader context, the analysis done in Sect. 3.1 lies within the variety of studies that seek to explore relationships between density of a biological population, on the one hand, and sizes of organisms in it. The studies include the famous Reineke's rule and '$-3/2$ self-thinning rule'. Beginning in the 1950s, abundant literature appeared on volume-density relations in plant populations (see, e.g., [13, 31, 34, 35, 48] and others).

For many decades, forest practitioners had known that the number of trees in a forest stand was inversely related to mean sizes of the trees. For example, 100 years ago Frothingham [4, pp. 19–21], wrote: "It is thus evident that trees of the same age and growing in the same situations may vary somewhat in the rate of growth, according to their nearness to one another. ...Too many trees result in slender-stemmed, slow-growing stands..." Reineke [29] elaborated the understanding mathematically to find a relationship between stand density (N) and mean quadratic diameter at breast height ($\bar{d}_q$) called Reineke's rule:

$$N = k_0 \cdot \bar{d}_q^{\,k_1},$$

where k_0 is a constant and k_1 denotes a slope of the rule.

Initially, Reineke's rule was assumed to be described by a universal exponent $k_1 = -1.605$ but it was later found that the exponent may be species specific. For example, species-specific exponents were determined by Pretzsch and Biber [27] for four major European tree species. Vospernik and Sterba [42] compared the competition-density rule and the self-thinning rule in the context of potential density in 15 Austrian tree species. They analyzed the relations between the rules and argued the self-thinning rule to be a marginal case of the competition-density relationship.

The research by Yoda et al. [48] published in 1963 is one of the many works associated with the $-3/2$ rule theory. On the example of a wide variety of herbaceous and woody plants they established that dense plant populations followed a straight line in double log coordinates of mean plant weight $\overline{W}$ against plant density N. That is, these variables were related to each other by

$$\overline{W} = m_0 \cdot N^{m_1}, \tag{3.20}$$

where m_0 is a constant and m_1 is the slope of the relation. The authors claimed that slope $m_1 = -3/2$ was rather constant among many plant species, ages, and growing conditions. One can note that the slope independence of age brings about a dynamic sense of the rule (3.20) making the relation a line of self-thinning in time. The constancy of ratio $3/2$ in the exponent was related to suppositions that the volume of organisms is approximately the cubic power of their linear dimensions whereas organisms such as plants occupy an area on the ground that is proportional to the second power of the linear dimensions. The constancy of the slope m_1 should also imply that the organism proportions remain roughly the same, which in other words means geometric self-similarity.

Earlier, Hilmi [8] explored the same hypothesis as did Yoda et al. [48]—geometric similarity—but studied another relationship, between linear dimensions of trees and the stand density and got, naturally, the other slope of the self-thinning curve (−1/2).

Some authors comparing Reineke's rule with the −3/2 rule [24] explicitly give preference to Reineke's rule before the −3/2 rule, arguing the latter to be purely descriptive. Reineke's rule has been also used in modeling studies of growth and mortality. Palahí et al. [23] measured growth of Scots pine on permanent plots and their simulated growth. They found that the observed self-thinning slope was different from that predicted by Reineke's rule but was consistent among species as reported by other research.

Nevertheless, it is the −3/2 rule that has been recognized to be an important empirical generalization [14] and has received abundant response in publications (recent analyses may be found in [16, 25, 39] and other papers).

During the decades following the first publications by the authors of the −3/2 rule a substantial amount of data has been published, both in support of the rule and in opposition to it. White and Harper [45] confirmed the rule and extended it to several species. Kofman [14] expressed an opinion that the deviations from the −3/2 rule in published data were due to violation of the primary supposition made to derive the rule, that is, the hypothesis of geometrical similarity. A few years later, Newton and Smith [20] gave an illustrative example of this opinion. Comparing various data on self-thinning in black spruce they received two distinct groups of data. Some data were consistent with the self-thinning rule, whereas others had sufficiently deviating parameters from the expected ones. The authors concluded that allometric relationships including diameters of trees were at least partly responsible for these deviations. In a review LaBarbera [15] criticized both early attempts to call the self-thinning rule a "law" and the statistical methods to estimate the values of the self-thinning exponents and joined the opinion that the "exponent is not a constant but rather a variable, one describing specific aspects of the plants involved and the biological situation"[15, p. 107]. Zhang et al. [50] found that many popular statistical methods (including ordinary least squares, corrected ordinary least squares, and reduced major axis) are rather sensitive to the choice of datapoints, which may be a source of variation in estimated self-thinning slopes. Other methods (such as quantile regression, deterministic frontier function, and stochastic frontier function) were reported to have the potential to estimate the linear limiting boundary slope, with no subjective selection of data subsets being applied. Xue et al. [47] measured self-thinning slopes in stands of a cypress family tree, Chinese fir (*Cunninghamia lanceolata*). They reported the self-thinning exponent for the total tree mass to be equal to −1.05. On the other hand, the self-thinning exponents substantially varied among different tree organs equalling −1.46 for stems, −0.93 for branches, −0.96 for leaves, −1.35 for roots, and −1.28 for shoots. Only the exponent for stems was not significantly different from −3/2; all others, in absolute value, were significantly lower.

Pretzsch and Schütze urged that species-specific properties of woody plants should not be ignored "in favour of a questionable general scaling law" [28, p. 637]. Pretzsch also studied the consistency of the self-thinning rule against data from long-term plots of a number of tree species [25]. He found that the observed exponents significantly corresponded with the considered rules only in a minority of cases and concluded that the "hope for a consistent scaling law fades away." Later, in the monograph *Forest Dynamics, Growth and Yield* Pretzsch repeated this idea saying that "To obtain a better understanding of competitive mechanisms in forest stands, further research should clarify scaling rules for individual species rather than continuing to search for 'the ultimate law', which may be like hunting for a phantom" [26, p. 407]. Von Gadow [40, p. 364] also called generality of the much publicized self-thinning rules to be "nothing but a myth."

Theoretical grounds underlying the existence of the self-thinning rule phenomena have been analyzed many times. Based on the thinning experiments data, on the concepts of the self-thinning rule, on Reineke's stand density index, and Assmann's theory, Sterba [33] stressed the physical sense of the self-thinning lines that show the potential (maximal) density of forest stands.

West et al. [43, 44] developed a theoretical framework explaining scaling laws observed in vascular organisms. The scaling laws have been reported to include allometric exponents that are simple multiples of 1/4. In particular, the metabolic rate of the entire organism B scales as a 3/4 power of the organism mass W (i.e., $B \propto W^{3/4}$), which is also known under the name "Kleiber's Law" [21]. From this relation, a sort of self-thinning rule may be derived under the hypothesis that all the available living resources of a location of growth are being used and, in the case of a forest, the maximal stand density has been achieved. Mathematically, the hypothesis may be given as $N \cdot \overline{B} = c$ where c is a constant that shows the upper limit of the resource amount and $\overline{B}$ is the mean metabolic rate. Then, $\overline{B} = c \cdot N^{-1} \Rightarrow \overline{W} \propto c^{4/3} \cdot N^{-4/3}$, which suggests that the slope of the self-thinning line should be $-4/3$ rather than $-3/2$.

Larjavaara [16] elaborated a physiological framework based on consideration of toppling risk and maintenance cost in overcrowded populations at the state of stagnation equilibrium. The analysis of the ratio of inner bark and sapwood mass in the maintenance cost showed that the power exponent should be between $-3/2$ and $-2/3$ depending on the ratio. Vanclay and Sands [39] working with a dynamic model found that self-thinning may be regulated by the maximum basal area and in this case the slope equals -2. If the slope differs from -2 other additional predictors have to be used. Tausch [36] developed a model that takes into account species-specific features such as functional depth of foliage and the average mature size. The model analysis showed that the smaller the ratio between the functional depth and mature size the closer the forest stand was to a theoretical monolayer case producing a self-thinning slope of -2, whereas larger values of the ratio corresponded to a multilayer case that had a slope of -1.333.

In addition to the comparison of the self-thinning rule with empirical data, a number of efforts have been made to rethink the role and the status of the $-3/2$

rule. Lonsdale [17] came to the conclusion that by the time of his publication no evidence of the rule holding was available. He nevertheless admitted that the existence of an ideal limiting line could not be rejected until crucial experiments were performed. Hamilton et al. [7] commented that the size-density trajectories followed by plant populations under the self-thinning did not necessarily have a −3/2 slope. However, in their view, a more general assertion of some power rule giving an upper limit to self-thinning populations remained intact. Usoltsev [37] in a review of the discussions around the −3/2 rule noted that the edge of the criticism of the rule was directed at whether the slope of the self-thinning curve equalled −3/2. At the same time, according to him, no doubts have been expressed of the very existence of a line limiting the self-thinning. In general, the endeavors to establish a unified law reflect the human propensity to reduce complexity [19], although nature does not have to fulfill the wish. Still, as Matyssek et al. [19, p. 564] point out, "general allometry laws may be suitable for scaling when evaluating principles in self-thinning of woody versus herbaceous plant systems for their extent of similarity."

An analysis of the −3/2 rule performed by Zeide [49] is of special interest for self-thinning questions. Instead of the original formulation of the rule he considered the relationship linking the total biomass of trees with their number per unit area, which gives the slope −1/2. Analyzing the crown closure dynamics data from permanent sample plots Zeide found that when the forest canopy was dense the slopes were steeper than the suggested value −1/2 and when gaps accumulate in the canopy the slopes should be flatter than −1/2. Zeide came to the conclusion that the limiting line of self-thinning did not have a constant slope and the line should be curvilinear (in the log–log space). This suggestion of curvilinearity in the work by Zeide [49] is a rather seldom one in the self-thinning literature. Moreover, the curvilinearity is seen as inconvenient from a practical viewpoint for construction and use of density measurement diagrams [11], for example. However, it is necessary to admit that for a deeper understanding of the self-thinning phenomena the nature of the line should be analyzed independently of practical convenience. The form of the self-thinning trajectory in its full complexity should be theoretically comprehended. As Niklas et al. [22] pointed out, a notable feature of many analyses performed in the field of self-thinning is that they have been done in a posteriori style, that is, based largely on observations rather than on a profound mechanistic (theoretical) framework. Reynolds and Ford [30] have also urged the development of theoretical models that can explain how variations in the observed self-thinning slopes can occur.

The form of the self-thinning trajectory may be studied with simulation methods. Berger et al. [1] used a competition model KiWi that applied a "field-of-neighborhood" approach to take into account the strength of local interactions within a tree population in a mangrove forest. The authors found that the KiWi model is able to reproduce many features of self-thinning and they especially suggested that the simulated self-thinning curve may be divided into four stages on the plane "biomass-density". A linear section of the curve was interpreted to be a classical self-thinning line and was accompanied by the state of positive skewness in the stem diameter distribution.

Another side of the issue is that the empirical data appear in many cases as if a forest stand trajectory slowly approaches a straight line (in the log–log space) as if it were an asymptote. The evidence for that may be found in the work by Zeide [49] or, among recent observations, in a publication by von Gadow and Kotze [41]. Perhaps, because of such evidence the scientific discussion tended to give the rule an asymptote-like status, which was partially expressed as the limiting line idea. Smith and Hann [32] were probably the first to use the term "self-thinning asymptote" explicitly. Mathematically, one could not consider the line as a true asymptote because it would imply an infinite time of its approaching, therefore strictly speaking it might be seen as an asymptote-like "goal" of dynamics. Also, the line reflects a potential density of a forest stand [33].

The asymptote-like status of the rule means the admittance of a couple of theses: (1) the slopes of self-thinning trajectories may deviate from the $-3/2$ value (i.e., this value is not a marked out constant); and (2) in the space "size-density," a limiting line exists that serves as a "goal" for population dynamics which this dynamics tends to and, when reached, follows along the line.

It is also worth mentioning that the question of the self-thinning rule may be explored in two meanings, that are, however, not contradictory to each other. First, the densities of many forest stands could be related *statically* against their mean volumes/masses of trees. Second, the density of one particular forest stand could be followed *in time* as related to its mean volume/mass of trees.

In this section, the asymptote-like status of the $-3/2$ self-thinning rule is tested both theoretically and empirically. The self-thinning rule is understood in the second meaning, as a time development of a forest stand whereas static data sometimes have to be used to prove the analysis results.

3.2.1   Methods and Data

To test the status of the $-3/2$ rule a model of stem surface growth of a forest stand was explored, developed, and analyzed in Sect. 3.1. Despite the model's simplified (or one could even say "oversimplified") structure it seems to hold enough similarity with real forest stands to predict essential parameters of size–density and size–size relationships [5, 6]. Visually, the model corresponds to an idealized population of even-sized cones that can be described by the following relationships (see also p. 36)

$$r \propto \sqrt{\frac{1}{N^{\gamma_1}}} \quad \text{or} \quad \left(N^{-\frac{1}{2}\gamma_1} \right), \tag{3.21}$$

$$l \propto r^{\frac{2}{\gamma_2} - 1}, \tag{3.22}$$

where, as above, N is the quantity (or density) of figures, r and l are the radius of the cone base and generatrix, respectively, and γ_1 and γ_2 are parameters. To derive these relations it was supposed (see also Sect. 3.1.1) that (1) the cones are narrow and high enough to accept that the generatrix l is approximately equal to the cone height h, that is, $l \approx h$ and (2) if the quantity of the cones decreases their horizontal size grows accordingly.

As shown above (see p. 36), when the total stem surface $\hat{S}$ was independent of N, that is, $\hat{S}(N) = C = \text{const}$, then $\gamma_1 = \gamma_2$. The latter equality means that the power exponent in (3.21) can be directly predicted from (3.22) and vice versa. If $\hat{S}$ depends on N, that is, $\hat{S}(N)$ grows or decreases with the decrease in N (as a result of self-thinning), then relationships between γ_1 and γ_2 can be predicted.

Whittaker and Woodwell [46] have shown that the surface area of stem wood can be effectively estimated from linear regressions on the conic surface. Thus, the model provides an opportunity to formulate a simple and clear expression of the total stem surface area as (see also p. 33)

$$\hat{S} = \pi \cdot l \cdot r \cdot N. \tag{3.23}$$

The formula (3.23) along with (3.21) and (3.22) and the established properties of the model may be used to explore the question about the status of the −3/2 self-thinning rule.

To verify the results of the model analysis given below a number of data sources were used. The first group of data originates from levels-of-growing-stock studies in Douglas fir forest stands (see also p. 42). The cooperative research was begun in the early 1960s to study the relations between growing stock, growth, cumulative wood production, and tree size in repeatedly thinned stands. For analysis, five datasets were available: Hoskins [18], Iron Creek [2], Rocky Brook [3], Skykomish, and Clemons [12]. Each report, along with the data on experimental (thinned) plots, contained the results of measurements on three control plots that were not subjected to any treatment and therefore reflected a natural self-induced forest stand dynamics. It is these data on control plots that were used here. The following numbers of repeated measurements at different ages in the datasets were available for the analysis: Hoskins, 10 measurements (ages span from 20 to 55 years); Iron Creek, 10 (19–59 years); Rocky Brook, 7 (29–70 years); Skykomish, 9 (24–56 years); and Clemons, 9 (19–50 years). The reports contain abundant and detailed information on the Douglas fir stand development with many parameters including age of the plots, stand density N, mean height H, mean dbh D, quadratic mean diameter Dq, volume stock Vs, and others.

The total stem surface area BS for Douglas fir data was estimated using

$$BS = \frac{1}{2}\pi \cdot D_q \cdot H \cdot N. \tag{3.24}$$

that was found earlier to be a useful approximation to total stand surface area [9]. The use of quadratic mean diameter instead of mean dbh is quite important because the volume stock is calculated via the quadratic mean diameter in forest science. Therefore the measure has been used to provide compatibility with the data on the volume stock. It is the volume stock value that can be directly related to the −3/2 rule, which is in focus here.

The second group of data was extracted from the book by Usoltsev [38] that was used above (see p. 32). Among other parameters, the descriptions contain species mixture, site quality indices from I (the best) to V (the worst) according to the Russian scale of bonitation, age of forest stand, its density N, mean height H, mean dbh D, and volume stock Vs. The descriptions are grouped by Usoltsev [38] by the names of the authors and the following Scots pine (*Pinus sylvestris*) datasets were taken (the quantities of individual descriptions for each dataset are given in brackets):

1. Uspenskii, pine plantations in Tambov region, Russia, site index (SI) *Ia* (9), *I* (17), and *II* (4)
2. Mironenko, pine plantations in Tambov region, Russia, SI *Ia* (15) and *I* (7)
3. Kozhevnikov, pine plantations, Belorussia, SI *I* (5)
4. Gruk, pine plantations, Belorussia, SI *I* (7)
5. Gabeev, natural pine forests at Novosibirsk, Russia, SI *I* (5)
6. Kurbanov, natural pine forests at Yoshkar-Ola, Russia, SI *I* (6)
7. Heinsdorf, natural pine forest at Eberswalde, Germany, SI *Ia* (9), *I* (8), and *II* (11)

Each dataset for a definite site quality presents a number of descriptions of individual forest stands that differ by age. The data on individual forest stands were not subjected to grouping or aggregation and every individual description was treated as one measurement for a particular age. The latter means that the Scots pine data are static; that is, the data are descriptions of many forest stands at a particular age of the stands.

It should be mentioned that the model assumes changes in the state of a forest stand with the decrease of the stem density N. It is not "truly" dynamic as no account of time is involved but the model is also not static because it implies a single forest stand undergoing changes. Therefore the data are suitable for testing the model only to the extent to which the individual forest stands can be presented as a series of a single forest stand dynamics.

Because the model implies a consideration of a monotonic tendency of the total stem surface area a subsampling from the above-enumerated Scots pine datasets was sometimes necessary to extract such a simple monotonic tendency. As was done above (see p. 39) if a variation of total stem surface area was curvilinear in a dataset the following individual tendencies may be identified: increase, constancy, or decrease of the total stem surface area. Such a separation of different tendencies is of importance because simple power functions are used for modeling (see below in this section).

An advantage of the Scots pine datasets is that one can find there descriptions of the forest stands that are old enough to show sufficient decrease in measures of production, which is a rare case for data such as those of Douglas fir as most of the direct experiments operate with younger and hence well-growing forests. A consideration of the dynamics stages when forest stands experience decrease of production is also quite important for modeling.

The total bole surface area for the pine data was estimated by

$$BS = \frac{1}{2}\pi \cdot D \cdot H \cdot N. \tag{3.25}$$

It should be noted that the parameter of mean dbh is used in (3.25) because this is the only opportunity as no data on quadratic mean diameter are contained in the database by Usoltsev [38]. Because mean dbh is always numerically larger than quadratic mean diameter for the same stand data the use of quadratic mean diameter will, in principle, underestimate the value of stem surface area. On the other hand, this study does not assume comparison of the stem surface area between the Douglas fir and pine data; the estimated exponents are also for comparison only within the datasets. Therefore the difference of mean dbh and quadratic mean diameter should not influence inferences of the study.

The estimations of regression parameters for the studied relationships were performed with the help of STATISTICA 6 software. The software has a nonlinear estimation module that provides the tools to perform various regressions based on different loss functions. The user-specified regression model was mostly a two-parameter power function of the form $Y = a \cdot X^b$ where Y and X are the dependent and independent variables, respectively, and a and b are parameters. The main goal of using regression was estimation of the exponent b value.

At this point, an explanation is required regarding the goal of using the very form of the power function $Y = a \cdot X^b$. The power function is not as flexible as many other functions and from a statistical viewpoint it provides far from always best fittings for empirical data. For some empirical data, the power function would fail even to repeat the curvature of the data measurements, such as in Fig. 3.6.

Would this mean that the fitting by power functions reflected no properties of the data? From the standpoint of the goals of the study, the power function fitting does provide useful information of the data. In fact, the flatter the datapoints lie in Fig. 3.6 the closer the parameter b will be to zero, and the steeper the points lie the more

Fig. 3.6 An example of power function limitations to fit empirical data. Points represent some fictitious data

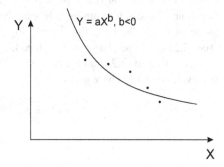

negative b will be. All this means that the value of b reflects important features of the relationship between X and Y independently of the low quality of the fitting. Such an approach is rather different from a traditional understanding of how a statistical fitting should be done. On the other hand, to perform an effective theoretical analysis not just only is information on the relationship between X and Y required, but the information should be in a definite mathematical form. As experience shows, the form of the power function is a very suitable one to perform a theoretical modeling in the field of forest structure.

3.2.2 Analysis of Model

Let us consider three types of relationships that describe the dynamics of an even-aged forest stand; these are dependencies of: (1) mean radius of tree trunks r on the stand density (i.e., $r(N)$); (2) total stem surface area $\hat{S}$ on the stand density (i.e., $\hat{S}(N)$); and (3) mean per tree volume stock $\bar{V}s$ on the stand density (i.e., $\bar{V}s(N)$). The relationship $r(N)$ according to the model takes the form (3.21). The functions $\hat{S}(N)$ and $\bar{V}s(N)$ are also supposed to take the form of a simple power function as

$$\hat{S} \propto N^{\alpha}, \tag{3.26}$$

$$\bar{V}s \propto N^{\beta}, \tag{3.27}$$

where α and β are exponents.

The relations (3.26) and (3.27) are not to say that the power function is the only or the best form to describe the relations. The grounds to use the power function were mentioned above. Technically, it is implied that real nonmonotonic curves may be divided into more or less monotonic segments within which a certain tendency can be seen, for example, fast growth, slow growth, slow decrease, and the like. Supposedly, it is these segments that may be approximated by a power function and the exponent of the function gives an estimation of the intensity of the tendency, growth or fall. As said above, the form of a power function allows one to study the relationships easily using a purely analytical approach.

One can note that the exponent β in (3.27) is the very factor that has been a keystone of the self-thinning rule's analysis because it corresponds exactly to the slope of a self-thinning line in log–log coordinates. The interrelationships of γ_1 (see (3.21)), α, and β are now explored.

According to (3.23) an expression for $\bar{V}s(N)$ may be found. A multiplication of both sides of (3.23) by r gives

$$r \cdot \hat{S} = \pi \cdot l \cdot r^2 \cdot N = \hat{V}s,$$

where $\hat{V}s$ stands for the total volume stock of the model population. Dividing the expression by N produces a basic formula for the mean per tree volume stock as

$$\bar{V}s = \frac{r \cdot \hat{S}}{N}. \tag{3.28}$$

Expression (3.28) may be analyzed by various methods because the relationships $r(N)$ and $\hat{S}(N)$ can be known or assessed. Let us take the relationships $r(N)$ and $\hat{S}(N)$ in the form (3.21) and (3.26), respectively. Then (3.28) can be rewritten as

$$\bar{V}s = K \cdot N^{-\frac{\gamma_1}{2}} \cdot N^{\alpha} \cdot N^{-1} = K \cdot N^{\alpha - \frac{\gamma_1}{2} - 1}, \tag{3.29}$$

where K is a normalization constant. It follows from (3.29) straightforwardly that

$$\beta = \alpha - \frac{\gamma_1}{2} - 1, \tag{3.30}$$

which permits a couple of inferences regarding the main question of the study, namely, which status the self-thinning rule should have.

The first inference seems as if *there cannot be one unique self-thinning line slope β that fits with every forest stand*. The reason to think so may be that both γ_1 and α have a high probability to be very individual for particular forest stands subjected to self-thinning. The parameters should depend on a number of internal and external factors such as spatial distribution of trunks, genetic variability of the population, and so on. There is yet an opportunity that the parameters γ_1 and α, varying individually in opposite directions, might each compensate the variations of the other. It is hard, however, to imagine that γ_1 and α from different forest stands are so finely tuned to each other to give the same β in all cases.

Another inference from (3.30) is that *even within one particular forest stand the slope β does not remain constant in the course of time*. Indeed, there is nothing forbidding both γ_1 and α from varying during the life of an individual forest stand. It may especially concern the latter parameter as at the growing stage of $\hat{S}(N)$ $\alpha < 0$, when $\hat{S}(N)$ remains constant $\alpha \approx 0$ and when $\hat{S}(N)$ falls $\alpha > 0$. Even if γ_1 alters correspondingly it is unlikely to expect γ_1 and α to compensate each other exactly. This can be easily demonstrated on the example of the most prolonged dataset on Douglas fir growth, the Hoskins experiment [18] (Fig. 3.7).

In Fig. 3.7a, the dynamics of the total stem surface area estimated through (3.24) was visually divided into more or less monotonic segments (marked by straight lines) that corresponded to the defined stages/tendencies of growth: fast growth, slow growth, and decrease. Then for the same stages in the relationship $\bar{V}s(N)$, the parameter β was estimated (Fig. 3.7b). It is clear from the example that the slope in $\bar{V}s(N)$ varies with the reduction in stand density, and more precisely, it changes from the value $\beta < -3/2$ to $\beta > -3/2$, which is dealt with below.

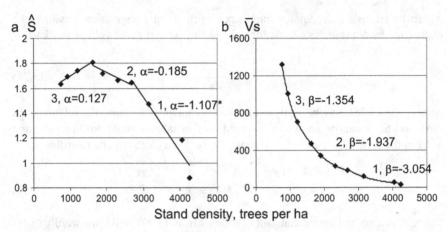

Fig. 3.7 Alterations of exponents in the relationships $\hat{S}(N) \propto N^\alpha$ (**a**) and $\bar{V}s(N) \propto N^\beta$ (**b**) in the Hoskins experiment [18] in the course of growth of the same forest stand. The $\hat{S}$ axis denotes the estimation of the total stem surface area according to (3.24) in ha. The $\bar{V}s$ axis denotes the mean per tree trunk volume in cubic decimeters. The *straight lines* in (**a**) serve merely to mark the defined stages of dynamics, key: *1*—fast growth, *2*—slow growth, *3*—decrease of $\hat{S}$. The values of β in (**b**) were estimated in the same stages of the relationship $\bar{V}s(N)$. With asterisk * the value is denoted that is significant at $p < 0.1$. All other estimated parameters are significant at $p < 0.05$. The values of γ_1 of the relationship $r(N)$ for the same defined stages are: (*1*) 2.114; (*2*) 1.277; (*3*) 1.025 (all are significant at $p < 0.05$)

3.2.3 Accuracy of the Model Prediction of the Slope in $\bar{V}s(N)$

In the above-given example, the values of β were estimated by regression procedures, thus they have an empirical status. At the same time, it is worth comparing the regression values of β with the values of β calculated through γ_1 and α to see how the formula (3.30) is good in principle.

To evaluate the formula (3.30) all the datasets used were treated as follows. The curves of the total stem surface area dynamics $\hat{S}(N)$ were divided into more or less monotonic segments, analogically with Fig. 3.7a, with each segment expressing a definite tendency of growth. Then the parameters γ_1 and α were estimated for the segments and with their help the calculated values of β_c were found using (3.30). For the same segments, the regression values of β_r were found. The comparisons of β_r and β_c are given in Tables 3.2 and 3.3 and in Fig. 3.8.

As follows from Table 3.2 and Fig. 3.8 there is sufficiently good correspondence between β_c calculated from (3.30) and β_r approximated directly from Douglas fir data. In most of the cases the model (i.e., β_c) slightly underestimates the values of β_r. Taking, however, into consideration that the forest stand structure model is a very simple one suggests that such a result may be more than satisfactory.

On the other hand, a comparison of β_c and β_r for the Scots pine data reveals larger deviations between them (Table 3.3). The most probable cause for the deviations is

Table 3.2 Estimated $(\gamma_1, \alpha, \beta_r)$ and calculated (β_c) parameters for the Douglas fir database

Dataset ID	TS[a]	$\gamma_1^\dagger$	$\alpha^\dagger$	$\beta_r^\dagger$	β_c	$\dfrac{\beta_c - \beta_r}{\beta_c}$, %
Hoskins	fg	2.114	−1.107*	−3.054	−3.164	3.49
Hoskins	sg	1.277	−0.185	−1.937	−1.823	−6.22
Hoskins	d	1.025	0.127	−1.354	−1.385	2.24
IronCreek	fg	4.116	−3.239	−5.484	−6.297	12.91
IronCreek	sg	1.210	−0.122*	−1.720	−1.727	0.41
RockyBrook	fg	4.555	−3.945	−6.731	−7.223	6.81
RockyBrook	sg	1.320	−0.237**	−1.761	−1.897	7.15
Skykomish	fg	29.729**	−30.113*	−42.746*	−45.977*	7.03
Skykomish	sg	2.016	−1.017	−2.895	−3.026	4.32
Clemons	fg	12.531*	−12.081	−18.074	−19.346	6.57
Clemons	sg	1.640	−0.634	−2.358	−2.454	3.90

*Significant at $p < 0.1$; **Not significant at $p < 0.1$; [†]Unmarked values are significant at $p < 0.05$
[a]TS = tendency of segment; it implies the slope of a monotone segment in the $\hat{S}(N)$ curve: fg = fast growth, sg = slow growth, d = decrease

Table 3.3 Estimated $(\gamma_1, \alpha, \beta_r)$ and calculated (β_c) parameters for the Scots pine database

Dataset ID	TS[a]	$\gamma_1^\dagger$	$\alpha^\dagger$	$\beta_r^\dagger$	β_c	$\dfrac{\beta_c - \beta_r}{\beta_c}$, %
Uspenskii I	fg	1.809	−0.972	−2.144	−2.877	25.45
Uspenskii I	fl	1.103	−0.005**	−1.425	−1.556	8.40
Uspenskii I	d	1.038	0.118	−1.333	−1.401	4.80
Uspenskii Ia	fl	1.105	−0.006**	−1.434	−1.558	7.97
Uspenskii Ia	d	1.037	0.118	−1.338	−1.400	4.47
Uspenskii II	d	1.045	0.100	−1.348	−1.422	5.20
Mironenko I	fl	1.229	−0.016**	−1.544	−1.630	5.30
Mironenko Ia	sg	1.267	−0.119*	−1.642	−1.753	6.31
Kozhevnikov	g	1.434	−0.380	−1.705	−2.097	18.67
Gruk	g	1.522	−0.719	−1.926	−2.480	22.32
Gabeev	d	1.151	0.154**	−1.230	−1.422	13.52
Heinsdorf I	d	1.017	0.182	−1.221	−1.326	7.95
Heinsdorf Ia	d	1.029	0.156	−1.225	−1.359	9.85
Heinsdorf Ia	shd	0.981	0.234	−1.189	−1.257	5.40
Heinsdorf II	sd	1.138	0.043	−1.429	−1.526	6.35
Heinsdorf II	shd	0.993	0.221	−1.197	−1.275	6.09
Kurbanov	d	0.775	0.360	−1.196	−1.027	−16.47

For initial data see Appendix Tables A.6, A.7, A.8, A.9, A.10, A.11, and A.12
*Significant at $p < 0.1$; **Not significant at $p < 0.1$; [†]Unmarked values are significant at $p < 0.05$
[a]TS = tendency of segment; it implies the slope of a monotone segment in the $\hat{S}(N)$ curve: fg = fast growth, fl = flat, g = growth, sg = slow growth, d = decrease, sd = slow decrease, shd = sharp decrease

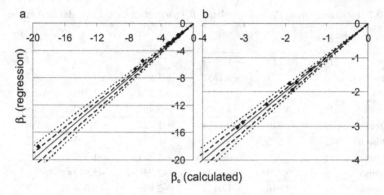

Fig. 3.8 Relation between figures calculated according to (3.30) (β_c) and estimated through regression (β_r) slopes of the relationship $\bar{V}s(N)$. (**a**) Range of β shown up to -20, (**b**) the upper right cluster of points is shown at a larger scale; values of β are up to -4. The *straight solid lines* denote $\beta_c = \beta_r$, and the *dashed lines* denote 5% and the *dotted lines* denote 10% deviations, respectively

that the values of mean dbh, not those of quadratic mean diameter, were used in the calculations of the total stem surface area according to (3.25). The quadratic mean diameter is known to be not only different from mean dbh but also related to it by a curvilinear relationship. This can be shown again on the example of the Hoskins experiment data. The quadratic mean diameter D_q measured consecutively over time in the Hoskins forest stand is related to the mean dbh D as $D_q = 0.012D^2 + 0.04D + 4.47$ ($R^2 = 0.9994$), with the constant term being significant at $p < 0.05$. Such a result implies that if $D = 0$ then $D_q \neq 0$, which may be a source of the deviations if the mean dbh is used instead of the quadratic mean diameter. It is necessary to note that, theoretically, $D = 0$ and $D_q = 0$ at the same time but for the real data located far from the zero reference point the relation between the diameters may introduce a sort of distortion in the calculations.

3.2.4 Relation of β_r to α in the Context of $-3/2$ Slope

The model says (see, e.g., (3.30)) that the value of α exerts a sufficient influence on the value of β. This is because, first, according to the observations α may be relatively large and, second, it may change its sign to the opposite. That is why it is worth looking at how estimated α relates to the estimated β_r.

Figure 3.9 summarizes the data on relationships $\beta_r(\alpha)$ for the Douglas fir database. A couple of findings can be derived from the chart. In spite of the substantial differences in the initial stand densities (4244 trees/ha in Hoskins vs. 1467 trees/ha in Skykomish experiments; see Fig. 3.10) and their different development over time, the trajectories of all the Douglas fir stands go through a

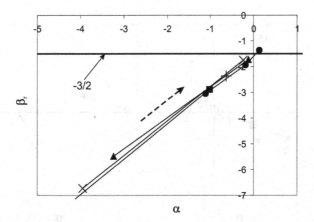

Fig. 3.9 Relationships between self-thinning slope β_r and the rate of total stem surface area growth α for Douglas fir database. Dataset key: *filled circle*—Hoskins, *filled triangle*—Iron Creek, *times symbol*—Rocky Brook, *filled square*—Skykomish, *plus*—Clemons. Each symbol corresponds to a row in Table 3.2 (i.e., a segment of the growth curves). The *solid lines* connect symbols denoting the same dataset. *Dashed arrow* shows the direction of time. That is, in terms of time, *lower left symbols* correspond to earlier moments and the *upper right symbols* correspond to later moments. The lower left points for Skykomish and Clemons are not shown as they lie far beyond the chart plane

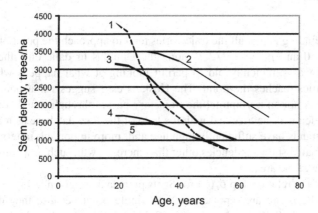

Fig. 3.10 Development of stem density in Douglas fir forest stands. Key: *1*—Hoskins, *2*—Rocky Brook, *3*—Iron Creek, *4*—Clemons, *5*—Skykomish. The graph was constructed by the author on the basis of tables published in [2, 3, 12, 18]

very narrow band in terms of the relationship $\beta_r(\alpha)$. This may imply that all the Douglas fir plantations show an extremely high similarity of growth on the plane "α–β_r."

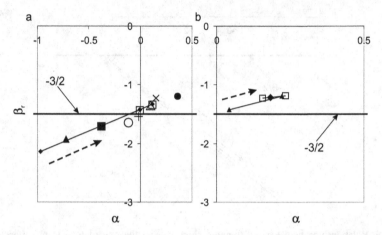

Fig. 3.11 Relationships between the self-thinning slope β_r and the rate of the total bole surface area growth α for Scots pine database. (**a**) *filled diamond*—Uspenskii I, *small open square*—Uspenskii Ia, *big open square*—Uspenskii II, *plus symbol*—Mironenko I, *open circle*—Mironenko Ia, *filled square*—Kozhevnikov, *filled triangle*—Gruk, *times symbol*—Gabeev, *filled circle*—Kurbanov; (**b**) *filled diamond*—Heinsdorf I, *small open square*—Heinsdorf Ia, *filled triangle*—Heinsdorf II. Each *symbol* corresponds to a row in Table 3.3 (i.e., a segment of the growth curves). The *solid lines* connect symbols denoting the same dataset. The *dashed arrows* show the direction of time dynamics

Another finding is that all the trajectories tend to approach the point of the chart where $\alpha = 0$ and $\beta_r = -3/2$. Among the Douglas fir data, only the Hoskins forest stand was sufficiently old to permit looking at what happens when forest stand dynamics reaches the point. The Hoskins case suggests that point $\alpha = 0$, $\beta_r = -3/2$ is not an asymptote; the trajectory passes through it and continues to grow. This idea may be assessed with the help of the pine database as many of the pine forest stands have sufficiently high age and, more important, in terms of total stem surface area show tendencies other than mere growth, such as a constant level or a decrease of the area.

Details of the relationship $\beta_r(\alpha)$ for Scots pine data are given in Fig. 3.11. Some datasets in the figure are represented by a single point because they describe a single tendency in $\hat{S}(N)$ development (see Table 3.3). Nevertheless, both multipoint trajectories and single points representing quite different pine forest stands tend to lie in a narrow band and cross the point $\alpha = 0$, $\beta_r = -3/2$, as well as in the case of Douglas fir. It is also clearly seen that the more positive values of α are the farther the values of β_r are up from the line $-3/2$. Especially, the examples of Heinsdorf Ia and Heinsdorf II datasets (see Table 3.3, Fig. 3.11b) show that the sharper the decrease of $\hat{S}(N)$ is the farther the value of β_r shifts up from the line $-3/2$.

There are therefore some inferences that follow from the data. The first is that forest stands in the course of their self-thinning tend to cross the point $\alpha = 0$, $\beta_r = -3/2$ on the plane α–β_r. Remember that $\alpha = 0$ means independence of total stem surface area $\hat{S}$ of N; that is, while the stand density proceeds to fall down the

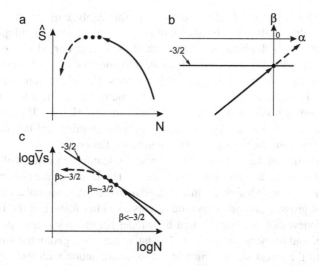

Fig. 3.12 Idealized descriptions of the main features of self-thinning processes: (**a**) development of the total stem surface area with the stand density $\hat{S}(N)$; (**b**) relationship between the power exponent α in $\hat{S}(N)$ and the power exponent β in the self-thinning curve $\bar{V}s(N)$; (**c**) the transition of β from $\beta < -3/2$ through $\beta = -3/2$ to $\beta > -3/2$. *Solid arrows* denote the stage when $\beta < -3/2$ and the total stem surface area increases. *Dots* designate the stage when $\beta = -3/2$ and the area stays constant. The *dashed arrows* indicate the stage when $\beta > -3/2$ and the area decreases

area $\hat{S}$ stays constant. Taking into account (3.30) shows that the only solution for $\alpha = 0$ and $\beta_r = -3/2$ at the same time is when $\gamma_1 = \gamma_2 = 1$. The latter equality and (3.22) imply that the height of a tree is proportional to the radius of its base; that is, geometrical similarity takes place over a certain span of the forest growth time. *This is the particular time when these three peculiarities coincide: (1) independence of the total bole surface area $\hat{S}$ of N, (2) geometrical similarity in the growth of trees, and hence (3) the value of the self-thinning slope $\beta_r = -3/2$.*

Another inference is that *the value −3/2 is not a goal (in the sense of an asymptote) for development of a self-thinning curve slope*. The value of β_r passes through the level of −3/2 and if α continues to grow then β_r grows correspondingly.

These ideas are presented graphically in Fig. 3.12.

As mentioned above, comparisons of the theoretically derived self-thinning rules with empirical data [25] show that the estimated slopes vary significantly among species and individual forest stands. The analysis performed in this section deals with variations of the slope within an individual forest stand and it reveals that the −3/2 rule has a definite and special place in these variations.

Application of the simple geometrical model to the Douglas fir and Scots pine data suggests that the slope of the self-thinning curve $\bar{V}s(N)$ will not remain constant during the course of growth and self-thinning of a single forest stand. Most probable, at the initial stages of stand growth the slope will be less than −3/2 and at old ages of the stand the slope will be higher than $-3/2$. Inevitably, a time will come when the slope exactly equals $-3/2$. In other words, the slope $-3/2$ is an obligatory state in the course of self-thinning of a forest stand. A similar consideration has been suggested by Zeide [49] who derived the development of the self-thinning curve slope from canopy closure dynamics. In this section, the self-thinning slopes were explored on the basis of a very transparent geometric model of a forest stand. It should also be mentioned that as early as 1970 White and Harper [45] admitted that older self-thinning stands tended to deviate from the limiting self-thinning line so that the last points of the trajectories lay on another, flatter, line.

At the very time of the $-3/2$ slope two particular features coincide with it. One is that the total stem surface area remains constant. The length of the constancy stage would probably vary with species, initial stand densities, their spatial arrangements, conditions of growth, and other specific factors. Another feature of the time, from the point of view of the model, is that the model parameters γ_1 and γ_2 are equal to each other and unity, $\gamma_1 = \gamma_2 = 1$. The latter implies a geometric similarity in the growth of the forest stand, which is not in contradiction with the $-3/2$ rule as formulated by its authors.

To put it briefly, the slope $-3/2$:

1. is a very specific and obligatory state in the process of forest stand growth and
2. is not an asymptote-like but rather a transitional point (or maybe a span) in the time of growth

These two assertions may be called a *transitional status* of the $-3/2$ rule.

An important issue should be noticed concerning similarity in forest stand functioning, especially in the context of self-thinning. As said above (see Figs. 3.9 and 3.11), both Douglas fir and Scots pine trajectories on the plane "$\alpha–\beta$" run within rather narrow bands of values. Do the trajectories, however, have the same unique slope or not? The formula (3.30) suggests that the slope in $\beta(\alpha)$ should be only one and namely equal to unity. It should be kept in mind that the theory applied provides sufficient transparency of analysis but may ignore some factors. A more detailed theory might, for example, consider a relationship $\beta \propto k \cdot \alpha$ where k is a normalization constant. This relationship does not lose important properties of (3.30) but may take into account species-specific factors.

In fact, Fig. 3.13 shows the slopes for Douglas fir and Scots pine data differ in comparison with the slope 1.0 and with each other. In the figure, linear regressions were computed for all the Douglas fir data, on the one hand, and all the Scots pine data. It is clear from the diagram that regression lines for both species are different from unity. It is important, however, to see that the slopes may be accepted as rather constant *within* the species. This means that the relationship $\beta(\alpha)$ remaining species specific may give evidence of a high similarity of stand development within tree species.

In general, introduction of stem surface area into analyses of forest stand structure and dynamics seems to be a rather fruitful approach that allows getting important inferences in the research scope [9, 16]. A geometric model of a forest

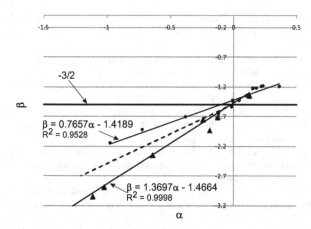

Fig. 3.13 A summarized presentation of Douglas fir and Scots pine data on the plane "α–β" where α is the power exponent in $\hat{S}(N)$ and β is the power exponent in $\bar{V}s(N)$. Key: *dashed line* shows the slope equal to unity; *solid lines* are linear regression lines for all the Scots pine (*filled circle*) and all the Douglas fir (*filled triangle*) data. For the sake of visibility, only the upper right points for Douglas fir are shown. The regression equations are shown on the diagram

stand using the approach of total stem surface area [5] has proved to be rather helpful at analyzing self-thinning processes in real forest stands. Despite its extreme simplicity the model appears to have enough similarity with real even-aged forests because the model's predictions are often reasonably close to measured values of power exponents.

References

1. Berger U, Hildenbrandt H, Grimm V (2002) Towards a standard for the individual-based modeling of plant populations: self-thinning and the field-of-neighborhood approach. Nat Res Model 15(1):39–54
2. Curtis RO, Marshall DD et al (2009) Levels-of-growing-stock cooperative study in Douglas-fir: Report No. 19–the Iron Creek study, 1966–2006. United States Department of Agriculture, Forest Service, Pacific Northwest Research Station
3. Curtis RO, Marshall DD et al (2006) Levels-of-growing-stock cooperative study in Douglas-fir: Report No. 18-Rocky Brook, 1963–2006. US Department of Agriculture, Forest Service, Pacific Northwest Research Station
4. Frothingham EH (1914) White pine under forest management. Bull Dep Agric 13:1–70
5. Gavrikov VL (2014) A simple theory to link bole surface area, stem density and average tree dimensions in a forest stand. Eur J For Res 133(6):1087–1094
6. Gavrikov VL (2015) An application of bole surface growth model: a transitional status of '−3/2' rule. Eur J For Res 134(4):715–724
7. Hamilton NRS, Matthew C, Lemair G (1995) In defense of the −3/2 boundary rule: a re-evaluation of self-thinning concepts and status. Ann Bot 76:569–577
8. Hilmi GF (1955) Biogeophysical theory and prognosis of forest self-thinning. Izd-vo AN SSSR, Moscow (in Russian)

9. Inoue A (2004) Relationships of stem surface area to other stem dimensions for Japanese cedar (*Cryptomeria japonica* D. Don) and Japanese cypress (*Chamaecyparis obtusa* Endl.) trees. J For Res 9(1):45–50
10. Inoue A (2009) Allometric model of the maximum size–density relationship between stem surface area and stand density. J For Res 14(5):268–275
11. Jack SB, Long JN (1996) Linkages between silviculture and ecology: an analysis of density management diagrams. For Ecol Manage 86(1):205–220
12. King JE, Marshall DD, Bell JF (2002) Levels-of-growing-stock cooperative study in Douglas-fir: Report No. 17-the Skykomish study, 1961–93; The Clemons study, 1963–1994. Pacific Northwest Research Station, USDA Forest Service
13. Kira T, Ogawa H, Sakazaki N (1953) Intraspecific competition among higher plants. i. Competition-yield-density interrelationship in regularly dispersed populations. J Inst Polytech Osaka City Univ Ser D 4:1–16
14. Kofman GB (1986) Growth and form of trees. Nauka, Novosibirsk (in Russian)
15. LaBarbera M (1989) Analyzing body size as a factor in ecology and evolution. Ann Rev Ecol Syst 20:97–117
16. Larjavaara M (2010) Maintenance cost, toppling risk and size of trees in a self-thinning stand. J Theor Biol 265(1):63–67
17. Lonsdale WM (1990) The self-thinning rule: dead or alive? Ecology 71(4):1373–1388
18. Marshall DD, Curtis RO (2001) Levels-of-growing-stock cooperative study in Douglas-fir: report no. 15-Hoskins: 1963–1998. United States Department of Agriculture, Forest Service
19. Matyssek R, Agerer R, Ernst D, Munch J-C, Osswald W, Pretzsch H, Priesack E, Schnyder H, Treutter D (2005) The plant's capacity in regulating resource demand. Plant Biol 7(6):560–580
20. Newton PF, Smith VG (1990) Reformulated self-thinning exponents as applied to black spruce. Can J For Res 20(7):887–893
21. Niklas KJ, Kutschera U (2015) Kleiber's Law: How the Fire of Life ignited debate, fueled theory, and neglected plants as model organisms. Plant Signal Behav 10(7):e1036216
22. Niklas KJ, Midgley JJ, Enquist BJ (2003) A general model for mass–growth–density relations across tree-dominated communities. Evol Ecol Res 5(3):459–468
23. Palahí M, Pukkala T, Miina J, Montero G (2003) Individual-tree growth and mortality models for scots pine (*Pinus sylvestris* l.) in north-east Spain. Ann For Sci 60(1):1–10
24. Perala D, Leary R, Cieszewski C (1999) Self-thinning and stockability of the circumboreal aspens (*Populus tremuloides* Michx., and *P. tremula* L.). Research Paper NC-335. U.S. Department of Agriculture, Forest Service, North Central Research Station., St. Paul
25. Pretzsch H (2006) Species-specific allometric scaling under self-thinning: evidence from long-term plots in forest stands. Oecologia 146(4):572–583
26. Pretzsch H (2009) Forest dynamics, growth, and yield. Springer, Berlin/Heidelberg
27. Pretzsch H, Biber P (2005) A re-evaluation of Reineke's rule and stand density index. For Sci 51(4):304–320
28. Pretzsch H, Schütze G (2005) Crown allometry and growing space efficiency of Norway spruce (*Picea abies* [L.] Karst.) and European beech (*Fagus sylvatica* L.) in pure and mixed stands. Plant Biol 7(6):628–639
29. Reineke LH (1933) Perfecting a stand-density index for even-aged forests. J Agric Res 46(7):627–638
30. Reynolds JH, Ford ED (2005) Improving competition representation in theoretical models of self-thinning: a critical review. J Ecol 93(2):362–372
31. Shinozaki K, Kira T (1956) Intraspecific competition among higher plants. vii. logistic theory of the cd effect. J Inst Polytech Osaka City Univ D 7:35–72
32. Smith N, Hann DW (1984) A new analytical model based on the −3/2 power rule of self-thinning. Can J For Res 14(5):605–609
33. Sterba H (1987) Estimating potential density from thinning experiments and inventory data. For Sci 33(4):1022–1034
34. Sterba H, Monserud RA (1993) The maximum density concept applied to uneven-aged mixed-species stands. For Sci 39(3):432–452

35. Sterba H, Monserud RA (1995) Potential volume yield for mixed-species Douglas-fir stands in the northern Rocky Mountains. For Sci 41(3):531–545
36. Tausch RJ (2015) A structurally based analytic model of growth and biomass dynamics in single species stands of conifers. Nat Res Model 28(3):289–320
37. Usoltsev VA (2003) Forest phytomass of Northern Eurasia: limited productivity and geography. Ural Branch of Russian Academy of Sciences, Yekaterinburg (in Russian)
38. Usoltsev VA (2010) Eurasian forest biomass and primary production data. Ural Branch of Russian Academy of Sciences, Yekaterinburg (in Russian)
39. Vanclay JK, Sands PJ (2009) Calibrating the self-thinning frontier. For Ecol Manage 259(1):81–85
40. von Gadow K (1986) Observations on self-thinning in pine plantations. S Afr J Sci 82:364–368
41. von Gadow K, Kotze H (2014) Tree survival and maximum density of planted forests–observations from South African spacing studies. For Ecosyst 1(1):1–9
42. Vospernik S, Sterba H (2014) Do competition-density rule and self-thinning rule agree? Ann For Sci:1–12
43. West GB, Brown JH, Enquist BJ (1997) A general model for the origin of allometric scaling laws in biology. Science 276(5309):122–126
44. West GB, Brown JH, Enquist BJ (1999) A general model for the structure and allometry of plant vascular systems. Nature 400(6745):664–667
45. White J, Harper JI (1970) Correlated changes in plant size and number of plant populations. J Ecol 64:467–485
46. Whittaker R, Woodwell G (1967) Surface area relations of woody plants and forest communities. Am J Bot:931–939
47. Xue L, Hou X, Li Q, Hao Y (2015) Self-thinning lines and allometric relation in Chinese fir (*Cunninghamia lanceolata*) stands. J For Res 26(2):281–290
48. Yoda K, Kira T, Ogawa H, Hozumi K (1963) Intraspecific competition among higher plants. xi. Self-thinning in over-crowded pure stands under cultivated and natural conditions. J Biol Osaka City Univ 14:107–129
49. Zeide B (1987) Analysis of the 3/2 power law of self-thinning. For Sci 33(2):517–537
50. Zhang L, Bi H, Gove JH, Heath LS (2005) A comparison of alternative methods for estimating the self-thinning boundary line. Can J For Res 35(6):1507–1514

Chapter 4
Stem Respiratory Rate and Stem Surface Area

Abstract Analysis of biological phenomena often implies a clear distinction between *structure* and *function* although they are tightly interrelated. Structure is seen as a morphological basis of function. In this chapter, a popular biological function—respiration in trees—is reviewed as applied to the structure that is the focus of the book, the stem surface area of trees.

Keywords Growth respiration • Isometrical scaling • Maintenance respiration • Respiration • Stem surface area • Trees

4.1 Stem Surface Area and Other Measures in the Context of Respiration

Respiration is one of the most fundamental functions of living organisms that performs a transformation of organic matter in the organisms' cells. Its quite complicated molecular mechanisms have been consecutively deciphered by generations of scientists in the course of the twentieth century. A respiration chapter has nowadays become a standard part of any physiology or biochemistry textbook (see, e.g., [25, 30]).

Some molecular respiration mechanisms are remarkably stable in a vast majority of organisms: glycolysis is found in bacteria and in plant as well as animal cells. Other mechanisms may vary depending on whether respiration occurs under aerobic or anaerobic conditions and on the initial organic compounds involved. It is, however, known that respiration is basically a decomposition of complex carbohydrates, fats, and proteins into simpler compounds, with energy being extracted from the decomposition, which is necessary for cell functioning. Aerobic respiration, the most energetically effective, oxidizes carbohydrates into water and carbon dioxide.

© The Author(s) 2017
V.L. Gavrikov, *Stem Surface Area in Modeling of Forest Stands*,
SpringerBriefs in Plant Science, DOI 10.1007/978-3-319-52449-8_4

Since the 1950s [22], relatively easy measurements of CO_2 fluxes through infrared gas analyzers have been made possible. This allowed forest researchers to estimate carbon dioxide release, attributed to respiration, from different parts of trees.

Respiration is an intracellular process and the features and amount of respiration may be studied at various scales, from cell to ecosystem. Rayment et al. [31] measured gas exchange in a mature *Picea mariana* forest stand. The authors applied an approach of comparing photosynthesis and respiration parameters at scales of shoot, branch, and canopy. Working at such contrasting scales one has to use quite different instrument techniques, a minicuvette system for shoots and an eddy covariance flux measuring station for the canopy. The authors recorded small differences of photosynthetic parameters among the scales, which they interpreted as a high optimality of photosynthetic capacity for carbon uptake. At the same time, recorded nighttime respiration significantly increased in the sequence shoot-branch-canopy, which was attributed by the authors to the costs of supporting an increased amount of biomass.

Korzuhin et al. [21] used the approach of different scales but with other goals. The aim was to find minimal levels of primary production as a function of meteorological factors that ensure survival of a tree species. Respiration is one of key terms of carbon balance and the authors considered the balance at the scales of leaf, tree, and canopy in frames of an ecophysiological model. At leaf scale, the respiration term included daytime and nighttime respiration. Respiration costs for trees included leaf respiration, fine root respiration, growth respiration, and maintenance respiration. Under a definite set of climatic parameters, a tree species can only survive if the trees can yearly restore a necessary minimum of leaves and conducting tissues, which requires respiration. The authors computed a leaf creation cost parameter for a number of Eurasian tree species and found that the predicted geographic ranges for the species satisfactorily corresponded to the observed ranges.

In many tree respiration studies, the task was not to measure isolated CO_2 flux but to uncover relations of the fluxes to some morphological dimensions of the trees. The importance of such relations was apparently based on an idea that morphological measures of trees and stands can be acquired quite easily and on a much broader scale than gas flux estimations from high-end analyzers that can be applied rather locally, both in time and space. The subsequent use of the relations is therefore much a question of calibration and scaling; it is often desirable to extend a small number of local measurements to larger objects and areas, which may help to develop carbon balance sheets for plant communities and whole ecosystems. To perform the extension, one has to relate gas efflux measurements to morphological dimensions of different tree tissues.

Attempts to quantify the relations between CO_2 efflux and stem and branch dimensions have a rather long history. Kinerson [20] measured CO_2 release from stems, branches, and roots in loblolly pine (*Pinus taeda*) trees. Stem respiration was measured with the help of a chamber attached to the stem and branch and root respiration was measured on excised tissue fragments. As a first step, the author established power regression links between diameters of stems/branches and their corresponding surface areas and then regressed the diameters as predictors of respiration rates, also through a power function. It was found that the stem diameter predicts stem respiration during active growth at the relation strength of $R^2 = 0.48$ and the same relation for dormant time had the strength of $R^2 = 0.23$. Branch respiration was related to branch diameter at the strength of $R^2 = 0.54$. Coupling the data with the results of foliage and root respiration as well as with the dependence of respiration on temperature, Kinerson [20] suggested estimations of yearly respiration rate of the whole plant community. The estimations amounted to 682, 257, 994, and $60 \, g \, C \, m^{-2}$ for leaf, stem, branch, and root components, respectively. It is worthwhile mentioning that according to the author's findings the rate of respiration decreased with increasing stem diameter and a proposed cause of it was a decrease in living cell proportion with increasing stem diameter.

As follows from Kinerson's [20] work the author adhered to the opinion that respiration had to relate to the surface area of the stem or branch. Another assumption may be that respiration should be a function of total tree biomass, or at least of sapwood biomass. Sprugel [39] criticized the two assumptions claiming that they contradicted each other because surface area and sapwood volume are not necessarily well correlated. Sprugel expressed the opinion that authors of many preceding works failed to distinguish between different partitions of respiration. The partitions are *growth respiration* which ensures building of new tissues and *maintenance respiration* which provides energy to continue the living conditions of cells and tissues even without any growth (for a functional modeling of respiration see, e.g., [1]). Partitioning the two components of respiration was viewed by Sprugel [39] to be vital for success in relating measured respiration values and amounts of underlying morphological structures or physiological factors. Later, the issue of partitioning total respiration was discussed in detail in Sprugel et al. [40] where the authors urged that growth (construction) and maintenance respiration should be clearly distinguished.

Sprugel [39] reported respiration measurements in young trees of *Abies amabilis* that were aimed at revealing within—and between—tree variations in respiration. The methods included the use of chambers temporarily attached to stem and branch sections in various parts of the trees in the course of one growing season.

A measured value of carbon dioxide efflux is always a representation of total respiration including all its supposed components. Two different approaches were applied by Sprugel [39] to partition the components of respiration. Method 1 was a "subtraction" method that was based on the assumption that growth respiration at the end of the growing season was zero, thus the recorded efflux values could be entirely attributed to maintenance respiration. Then, assuming maintenance respiration to be constant, maintenance respiration can be subtracted from the efflux values recorded in the growing season to get estimations of growth respiration. The partitioned components can then be regressed against morphological parameters. Method 2, a "multiple regression method," used total respiration as a dependent variable and regressed it against a linear combination of morphological parameters assumed to represent growth or, respectively, maintenance respiration. Statistical significance of coefficients in the linear combination was then interpreted to be that the corresponding respiration component significantly contributed to the total respiration.

Applying method 1, Sprugel [39] found that September respiration (assumed maintenance respiration) was most closely linked by a linear function to sapwood volume ($R^2 = 0.72$), which showed that almost all respiration at the time of no active growth was determined by sapwood. According to the methodology, a subtraction of maintenance respiration from summer respiration values should reflect growth respiration. In the study by Sprugel [39], the obtained growth respiration values correlated in the best way with the current ring volume, especially at the time of fastest growth ($R^2 = 0.68$) but a significant correlation with sapwood volume or cambial surface was never recorded.

Results obtained by Sprugel [39] with the help of multiple regression (method 2) supported the findings by method 1. That is, current ring volume and sapwood volume were significant predictors of total segment respiration for most of the sampling dates, especially for the time of fast growth. For most of the dates, the influence of cambial surface was found to be indistinguishable from measurement error.

These results might remain in a sort of contradiction with morphological data of physical dimensions of stem sections used for measurements. The author reported values of correlations of sapwood:ring and ring:surface to be 0.77 and 0.76, respectively [39, Table 1]. It was not explained how it could be that ring volume was a good predictor of growth respiration but cambial surface, fairly well correlated with ring volume, was not.

A way to interpret the results by Sprugel [39] may come from a geometrical consideration. Let fragments of stems and branches be ideal cylinders and a fragment have the measures: radius of stem at the beginning of the growing season R, thickness of the current ring Δr, and the length of the fragment l (see Fig. 4.1). Then ring volume V_R may be expressed as

$$V_R = (\pi(R + \Delta r)^2 - \pi R^2) \cdot l = (2\pi R \Delta r + \pi \Delta r^2) \cdot l.$$

Fig. 4.1 Fragments of stems as ideal geometrical figures: R is bole radius at the beginning of the growth season; Δr is width of the current ring; l is the length of stem fragment

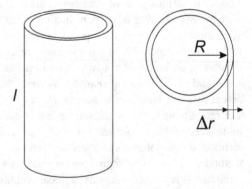

In the formula, neglecting the term containing Δr^2 as of the next infinitesimal order gives an approximate value of ring volume as

$$V_R \approx 2\pi R \Delta r \cdot l. \tag{4.1}$$

For the same stem fragment, cambial surface area is given as

$$S_C = 2\pi (R + \Delta r) \cdot l. \tag{4.2}$$

The values V_R and S_C may be compared with each other by dividing S_C by V_R, which gives

$$\frac{S_C}{V_R} = \frac{R + \Delta r}{R \Delta r} = \frac{1}{\Delta r} + \frac{1}{R}. \tag{4.3}$$

It is clear that the term $1/R$ in relation (4.3) gets smaller and smaller as the stem grows larger. For the case of larger stems therefore the relation (4.3) tends to $1/\Delta r$ as $\dfrac{S_C}{V_R} \to \dfrac{1}{\Delta r}$, which means that

$$S_C \cdot \Delta r \approx V_R. \tag{4.4}$$

Expression (4.4) may be interpreted so that everything depends on the variation of ring growth in the relation between current ring volume and cambial surface. If the ring growth is rather stable, that is, Δr varies insignificantly from year to year and may be considered to be a constant, then the correlation between ring volume and cambial surface should be rather strong. Therefore they should be equally good predictors of growth respiration. On the contrary, if Δr varies significantly among the years and no constant value of it may be accepted then the correlation between ring volume and cambial surface should be weaker. A question of which is a better predictor of growth respiration may be considered hypothetically but measurements are necessary for the final answer. The described geometrical approach would hardly work for thin branches because a number of suppositions (4.1) were used to get (4.4) but for larger stems the approach should be a helpful theoretical instrument to interpret obtained measurements.

The difference between current ring volume and cambial surface as predictors of growth respiration found by Sprugel [39] might be partly because thin stems with wide rings were the objects of measurements. On the other hand, cambial surface, that is, the amount of cells that can grow, reflects a *potential* growth, whereas the volume of the current ring is the *actual* growth that has occurred. From this standpoint, it is more logical to expect the ring volume to best correlate with growth respiration, which was demonstrated by Sprugel [39].

There may, however, be a sort of weakness in such results. One motive to correlate respiration measurements with morphological parameters is the hope of using morphological parameters subsequently as predictors of respiration. If thinking of measurements of trees on a mass scale, not to mention remote-sensing techniques, estimation of ring volumes might not be very much easier than direct efflux measurement. A more readily accessible parameter such as stem surface area therefore should not be discarded, at least for the case of larger stems for which current ring volume may well correlate with cambial surface (see (4.4)).

Sprugel [39] also suggested an approach to reconcile the premises of antecedent authors of respiration constancy per unit area and the observed wide variations of recorded respiration values. The first source of variations is variability in growth activity among individual trees, so that individual trees differ in growth respiration. The second source is differences of sapwood volume among trees and the fact that maintenance respiration is largely sapwood respiration. At the same time, Sprugel [39] found that total respiration per unit area along the stem was rather constant because growth and maintenance respiration balanced each other.

At this point, it would be worthwhile to note that respiration includes not only carbon dioxide emission but also oxygen consumption. Soloviev [38] experimented with excised fragments of pedunculate oak, (*Querqus robur*), stems. He found that 88–92% of oxygen consumption occurs in phloem, cambium, and the last tree ring and the balance is the consumption of deeper sapwood.

An example of when morphological parameters were used to estimate respiration rather than to measure it directly may be given by the study by Ryan and Waring [35]. The authors explored a question of whether maintenance respiration of stems can explain a decrease of net primary production observed in old-growth forests. A widespread idea of the sources of such a decrease considered a cessation of leaf index growth after closure of the canopy with continued growth of respiring woody biomass, which should give rise to increased maintenance respiration costs relative to the constant rate of photosynthesis and therefore to decreased net primary production. Ryan and Waring [35] measured wood production and sapwood volume in lodgepole pine stands aged 40, 65, and 245 years old. Sapwood volume was found through measurements of cores taken from the stems. Temperature-corrected maintenance respiration was estimated from sapwood volume via a nonlinear equation established in a previous study. Growth (construction) respiration was taken by the authors as 28% of wood production. As a result, construction respiration was $210 \, \text{g} \, \text{m}^{-2} \, \text{year}^{-1}$ in a 40-year-old stand and only $46 \, \text{g} \, \text{m}^{-2} \, \text{year}^{-1}$ in a 245-year-old stand. Maintenance respiration was estimated in the 40-year-old stand to be 61 and $79 \, \text{g} \, \text{m}^{-2} \, \text{year}^{-1}$ in the 245-year-old stand. Ryan and Waring [35] considered such an increase in maintenance respiration to be insignificant and concluded that growth of maintenance respiration cannot explain the decrease in net primary production in the 245-year-old stand. In their view, a more plausible cause of low wood production was low photosynthesis.

García [12] has also criticized the arguments by earlier authors according to which gross volume increment decreased with age because respiration losses should be roughly proportional to the surface area of stems and branches that was said to increase with age. Then growth should decrease with the product of mean diameter, height, and number of trees. In fact, as followed from data by García [12] for radiata pine the gross volume increment was approximately constant with age.

Ryan et al. [36] extended the study of maintenance respiration to various tissues of six pine species aiming at a comparison of physiologically close species that grew in different climates. According to the measurements, woody tissue respiration of large dormant stems was found to be low compared with other tissues, not exceeding $10 \, \text{nmol} \, C \, (\text{mol} \, C \, \text{biomass})^{-1} \, s^{-1}$. The annual total respiration for the pine species was found to be in the range 32–64% of carbon fixed annually. The ratio of annual respiration to photosynthesis was found to increase linearly with the biomass of younger pine stands. In another work, Ryan et al. [37] compared maintenance respiration in a number of conifer species. The authors measured autumn CO_2 release from stems of ponderosa pine, western hemlock, red pine, and slash pine. As in the previous citation, heartwood and sapwood parameters were measured from increment cores. Ryan et al. [37] established that maintenance respiration was linearly related to sapwood volume for all the species, with the regression slopes for pines being rather similar and that of hemlock differing from the pines. Temperature-corrected (at $10 \, °C$) respiration per unit of sapwood volume ranged from 4.8 to $8.3 \, \mu\text{mol} \, CO_2 \, m^{-3} \, s^{-1}$. The annual respiration fraction of carbon assimilated in photosynthesis was estimated by the authors to be 5–13% or, in physical units, $52–162 \, g \, m^{-2} \, \text{year}^{-1}$. Also, the fraction increased linearly with mean annual site temperature.

Scaling is one of the most important fundamental tasks in science. Probably, because of this "a relatively high proportion have been aimed at stand-level estimates" [40, p. 270] among studies on stem and branch respiration. As follows from considerations by Sprugel et al. [40] the key point in scaling respiration measurements on stems up to the stand level is a correct partitioning of respiration into construction and maintenance components. The importance of such a partitioning follows because respiration components mainly occur in *different morphological tissues*. To a great extent, maintenance respiration is a function of sapwood, thus to get the respiration component on the stand level one has to have equations to estimate the stand sapwood volume. The authors suggested generating species-specific equations linking sapwood volume, on the one hand, and height and sapwood area. Sapwood area was supposed to be an index of leaf area. Similarly, to get the stand-level values of construction respiration one first has to have estimations of respiration costs of production of different kinds of tissues. It was thus suggested to estimate the construction respiration for a stand simply from the annual wood production which is a routine measurement in studies of forest net primary productivity [40].

One of the goals of the study by Damesin et al. [10] was estimation of annual carbon flux from stems and branches in a *Fagus sylvatica* forest scaled up to stand level. The authors admitted that the scaling up involved sufficient difficulties inasmuch as initial measurements were mostly of local and noncontinuous nature whereas the desirable estimations were on an ecosystem level for the duration of the whole year. The results by Damesin et al. received in chambers generally corresponded to findings of other research about a linear relationship between maintenance respiration and sapwood. They mentioned, however, that maintenance respiration measured on a volume basis was higher in small suppressed trees than in others, which might be explained by different proportions between phloem and xylem. Also, a linear relationship was found between annual growth respiration and annual increment of stems. The authors found that area-based respiration steadily increased with stem diameter, which was especially well pronounced for growth respiration. The estimations by Damesin et al. [10] have shown that stem and branch respiration was a major component of ecosystem loss of carbon representing about 1/3 of the total flux.

In 2000 Amthor [2] published a comprehensive review of respiration studies over 30 years of their development. The focus of the review was on evolution of research paradigms (i.e., theoretical frameworks on which research is based) beginning with the simplest one which suggested explicitly differentiating between growth respiration and maintenance respiration. Later, it was recognized that some quantity of energy gained in respiration may be used neither for growth nor for maintenance; that is, it is simply wasted. At last, a general paradigm recognized that there were relationships between respiration and each biochemical process supported by respiration. Over decades, these study paradigms adopted in a wide circle of researchers played an important role in the scientific area stimulating the researchers to quantify links between the components of respiration and many individual physiological and biochemical processes.

It can, however, be noted that the review by Amthor [2] represents only one research approach, a mechanistic one. Amthor makes a sharp differentiation between mechanistic and empirical treatment of the topic. An empirical approach was said to describe data but to offer no explanation. According to this view, no information beyond the data is expected from the treatment. By contrast, a mechanistic approach explains data through processes known on lower levels of biological organization. For example, physiology is explained from biochemistry and biochemistry from chemistry and on and on down.

In fact, Amthor [2] consistently repeats that respiration should be related to physiological and biochemical *processes*, not morphological or anatomical *structures*. According to the author, this is "the appropriate approach for explaining respiration rates (or amounts), and is in contrast to simple empirical relationships between respiration and factors such as temperature and plant dry mass or surface area" [2, p. 13]. It is clear that the empirical approach is treated here as "less valuable."

At the same time, Amthor [2] had to admit that it was difficult to measure respiration support of growth and maintenance and improved measurements were needed. It is quite natural that there will always be difficulties in measuring within a mechanistic approach because a researcher will always need to explain a phenomenon through a lower-level phenomenon and to explain the found explanation through the next lower-level phenomenon and again and again downward. An achieved understanding will always demand a deeper understanding.

Having all this in view, a word in defense of the empirical approach should be and has been said. Amthor [2] himself cited Evans [11, p. 266] who wrote:

Selection for greater yield potential has not, could not and never shall wait on our fuller understanding of its functional basis, despite the pleas of physiologists... In that sense crop physiology may be *retrospective*, but the purpose of its *backwards* glance is to discern some of the ways *forward*... We have seen that, for example, the predominant improvements so far have not been in the efficiency of the major metabolic and assimilatory processes, but in the patterns of partitioning and the timing of development (italics added).

Undoubtedly, the greatest advantage of the mechanistic approach in respiration issues is its explanatory power. The advantage of the empirical approach focusing on structures and statistical relationships between them and measured gas exchange is, at least in lucky cases, a prediction that is badly needed by practitioners. If despite all the variations in measured variables of respiration a significant relationship has been established, it may be used for predictions.

A consideration of morphological and anatomical structures, not only processes, may be helpful in understanding what is observed. It has long been well known (see Sect. 1.1) that the vascular cambium produces secondary xylem and secondary phloem but there are important details in the process that are relevant to the respiration issue. To the inner side, an initial cambium cell produces derivative cells that sooner or later transform into mature vessels or tracheids. The initial cambium cell is, however, not alone in doing this work. In most plants, immediate descendant cells—immature xylem cells—continue to divide forming a more or less wide cambial zone [5, 34] which is, seen three-dimensionally, a volume of dividing and growing cells. Periclinal divisions of a cambium initial and its immature derivatives result in files of mature derivative cells that are especially characteristic of conifers [5, 34]. If immature xylem cells can preserve the ability to divide for a longer time span they can form quite a substantial volume of activity [43]. From this perspective, it is understandable that the growth respiration should be related mostly to the volume of forming tree rings as found by Sprugel [39]. Of course, the growth respiration of cambial initials per se does exist but may be inaccessible on the background of a larger amount of immature derivative growth respiration.

4.2 Respiration Versus Stem Surface Area: A Test of Isometrical Scaling Hypothesis

The theory developed by West et al. [41, 42] (known as WBE theory) provided the basis for understanding allometrical relationships in a wide range of organisms. Among other relationships they inferred that the metabolism B of an organism should scale to its body mass M with the power exponent 3/4, that is, $B \propto M^{3/4}$.

For smaller plants, WBE theory predicted an isometrical scaling, that is, $B \propto M^1$ [27].[1]

Direct measurements of respiration on the whole plant level are rather scarce because of obvious technical difficulties. Nevertheless recent studies have provided important data on whole plant respiration and its scaling across species and plant sizes [3, 4, 7, 8, 26, 32]. Reich et al. [32] reported measurements of about 500 seedlings and saplings belonging to 43 species that were both laboratory- and field-grown specimens and argued that the whole-plant respiration rate scaled approximately isometrically with total plant mass, which implies that the power exponent varies about unity. The authors also observed an isometric scaling of the whole-plant respiration rate to total nitrogen content, with this isometric scaling unaffected by various growth conditions including variation in light, nitrogen availability, temperature, and atmospheric CO_2 concentration.

Cheng et al. [7] showed that the aboveground respiration rates scaled as a 0.82-power of the biomass. In another study, Cheng et al. [8] reported the differences in scaling exponents for conifer and angiosperm tree seedlings: the scaling exponent for angiosperms was significantly higher (1.15) and for conifers lower (0.76) than unity. A combination of all the data in one dataset produced a scaling exponent of 1. Mori et al. [26] conducted extensive research of direct measuring of respiration in seedlings and large trees and established that the scaling allometric exponent varied continuously from unity for the smallest plants to 3/4 in larger saplings and trees.

Obviously, as the tree total size increases the total respiration of the tree body increases as well. However, the relative respiration per unit of body size may show at least two distinct behaviors as the body grows larger. Let R stand for the total respiration, γ for the scaling exponent, and V for the plant body volume. (For the sake of consistency, the plant volume V is used as a measure of the total body size assuming a good relationship between body volume and body mass [13].) Thus the scaling relationship is given by

$$R \propto V^\gamma. \tag{4.5}$$

If the scaling exponent γ is equal to unity then the per volume unit respiration should be constant, that is, independent of V, because $R/V \propto V^0$. However, if the scaling exponent γ is less than unity then per volume unit respiration cannot be a constant but should be a decreasing function of V:

$$\frac{R}{V} \propto V^{\gamma-1}.$$

It is widely understood that for larger trees the relative respiration per volume unit decreases with tree volume. Grounds for this decrease are that although the

---

[1]In the publication [13] a misprint occurred: the relation $B \propto M^1$ was incorrectly written as $B \propto M^0$.

bodies of smaller plants are metabolically active in the whole volume, low-active stem wood constitutes most of the biomass of larger trees [26, 28].

The same logic can be applied with respect to another measure of tree size, the stem surface area. Respiration, at least the construction part of it, of a tree stem is largely located in the thin sheath of inner bark [28, 29]. Unlike stem volume, an increase in stem surface area may result in increased metabolically active tissues. It may therefore be hypothesized that the relative respiration per area unit could be a constant independent of the size of the surface area.

Mathematically, the hypothesis is expressed as follows. Because the relationship between stem volume V and stem surface area S may be expressed through a scaling exponent β as

$$V \propto S^{\beta}, \tag{4.6}$$

then substituting Eq. (4.6) in (4.5) one gets the expression for the total respiration as

$$R \propto S^{\gamma\beta}.$$

Respectively, the relative respiration per area unit can be given as

$$\frac{R}{S} \propto S^{\gamma\beta-1}. \tag{4.7}$$

It is obvious that the scaling exponent $\gamma\beta - 1$ in relation (4.7) may be equal to zero, and for this to occur the relation between the scaling exponents β and γ should be as follows

$$\gamma = \frac{1}{\beta}. \tag{4.8}$$

To put it once again more precisely, it can be hypothesized that respiration per unit area in larger trees can be a constant, that is, independent of the amount of stem surface area. Formally, the hypothesis is presented in Eq. (4.8). Furthermore, one can note that a relation such as $R/S \propto S^0$ implies isometrical relationship of respiration R with stem surface S because increase of the surface by a unit would always add a constant value to the total amount of R. Thus the relation (4.8) could be read as a hypothesis of isometrical respiration scaling with stem surface. Such a hypothesis may be tested on the basis of available public data.

For the test, two kinds of data can be used: (1) values for γ, that is, for exponent scaling respiration to volume; and (2) values for β, that is, exponent scaling volume to stem surface.

The values for γ were taken from a number of sources [7, 8, 26, 32]. Additionally, the dataset published by Cheng et al. [7] was partly recalculated. The dataset contained dbhs, heights, log-transformed respiration, and log-transformed mass parameters for a number of species; among them are two conifers, *Pinus tabulaeformis* and *Pinus massoniana*. For these two conifers, a two-way volume equation provided by Inoue and Kurokawa [18] was applied to get bole volumes. The stem surface areas were estimated through a cone surface formula. The data for the two pine species were then fitted by a power function to get the scaling exponent β. Independently, the log-transformed respiration and mass measures were fitted by a linear function to get the exponent γ for the same combined pine data.

The values for β were taken from a study by Inoue and Nishizono [19] on the relationship between stem volume and stem surface area in Japanese cedar and Japanese cypress forest stands. The β values were also estimated from the datasets of levels-of-growing-stock studies in Douglas fir published by Marshall and Curtis [9, 23]. For control plots in the Douglas fir datasets, mean stem volumes were calculated by dividing stand volume by the number of living trees for every age available. From the same tables, stand diameters and heights were taken to estimate the mean stem surface areas using the cone formula. The mean volume was fitted against the mean stem surface area by a power function to estimate the β power exponent.

All the fittings were performed by means of STATISTICA 6 software using an ordinary least squares approach.

The values of γ and β scaling exponents, found in the literature and estimated, are summarized in Table 4.1. In most cases, if one of the two exponents was available the other was not. Due to limited data the unknown exponent was calculated with Eq. (4.8) and given in Table 4.1 in parentheses.

Table 4.1 Reported and estimated values of scaling exponents γ and β

Reported and estimated values of scaling exponents		
β	γ	References
$(1–1.33)^a$	From 1 to 0.75	Mori et al. [26]
(1.22)	0.82	Cheng et al. [7]
1.568 ± 0.032^b	0.7429 ± 0.032^b	Cheng et al. [7] (recalculated)
(1)	1	Reich et al. [32]
(0.87)	1.15	Cheng et al. [8], angiosperms
(1.32)	0.76	Cheng et al. [8], gymnosperms
(1)	1	Cheng et al. [8], angiosperms + gymnosperms
From 1.35 to 1.79	(0.74–0.56)	Inoue and Nishizono [19], Japanese cedar
From 1.35 to 1.63	(0.74–0.61)	Inoue and Nishizono [19], Japanese cypress
1.569 ± 0.006^b	(0.64)	Marshall and Curtis [23], Hoskins experiment, Douglas-fir
1.487 ± 0.007^b	(0.67)	Curtis and Marshall [9], Iron Creek experiment, Douglas-fir

[a]In parentheses, values estimated through Eq. (4.8) are given
[b]Fitted values $\pm$ std. error

According to Eq. (4.8), if the relative respiration per area unit were independent of the surface area then the scaling exponents γ and β should compensate each other so that their product would be equal to unity. An examination of the data in Table 4.1 shows that the measured values of the scaling exponent β tend to be slightly larger than those expected through Eq. (4.8). For example, the multispecies study by Mori et al. [26] gives the minimal value of γ as 0.75 which through Eq. (4.8) corresponds to the maximal value of β of ≈ 1.33. The study by Cheng et al. [7] gave a value 0.82 for γ which gives ≈ 1.22 for β. Cheng et al. [8] established that the scaling exponent γ for seedlings of two conifers was 0.76 which gives $\beta \approx 1.32$. The results for angiosperms suggest even larger γ values and therefore smaller β values (see Table 4.1). The actual measurements by Inoue and Nishizono [19], however, gave the values of β from 1.35 and larger.

In a study of aboveground respiration in mangroves, *Kandelia obovata* and *Bruguiera gymnorrhiza*, Hoque et al. [14, 16, 17] reported on relationships between respiration and tree sizes. The authors used a variable $D_{0.1H}^2 H$ as a proxy of stem volume where $D_{0.1H}$ is the stem diameter at 1/10 of height H. They found that a power function successfully approximated the relation of respiration to stem volume, with exponents from 3/4 (*K. obovata*) to 2/3 (*B. gymnorriza*). In other research Hoque et al. [15] studied seasonal variation in relationships of stem sizes and respiration in *Kandelia obovata* trees. Depending on the season, growing season or dormant winter, the exponents varied between 0.723 and 1.085, respectively. Also, respiration of small-sized trees reacted to the onset of seasonal growth more strongly than did large-sized trees. In the terminology used here, these exponents are γ exponents that correspond to β in the range between ≈ 1.33 and 1.5 (calculated with (4.8)).

As follows from the data, the values for γ tend to be slightly larger than those expected by Eq. (4.8). For example, the maximum γ values estimated by Eq. (4.8) for data by Inoue and Nishizono [19] and Hoskins and Iron Creek experiments [9, 23] amount to 0.74. The minimal γ value measured by Reich et al. [32], Cheng et al. [7, 8], and Mori et al. [26] was 0.75. All the comparisons mean that the product of γ and β appears to be slightly larger than unity.

Only in the case of data collected by Cheng et al. [7], was it possible to estimate the scaling exponents γ and β for the same dataset (see recalculation in Table 4.1). Multiplication of γ and β for the recalculated data gives $1.568 \times 0.7429 \approx 1.16$. Thus the relation in Eq. (4.7) for the data is given as $R/S = S^{0.16}$.

The estimations for γ and β values mean that the relative respiration per area unit should slightly increase with the increase in total stem surface area.

The inference may to some extent appear counterintuitive. In fact, if the properties of stem surface in the process of growth remained the same then there were insufficient causes for the relative respiration per area unit to alter. The data shown in Table 4.1 give evidence that in some, rather peculiar, cases the scaling exponents γ and β may satisfy Eq. (4.8) and therefore provide a certain stability of the respiration per unit area. Nevertheless it seems more likely that in general the scaling exponents γ and β do not satisfy Eq. (4.8) and they are expected to produce an increase of relative respiration per unit area as the tree stem grows larger.

A few hypotheses can be suggested to explain the increase in respiration per unit area. The allometric scaling concepts provide useful generalizations at the tree stem level. At the same time, the whole tree body has a complicated structure, which brings about a set of allometric relationships involving various tree body parts (for an analysis of the issue, see [27]). Sprugel [39] also showed that respiration per unit surface area varied strongly between sampling locations in young *Abies amabilis* trees. It is known that respiration varies in stems and branches of different diameters and branching orders [6]. Also, it has been shown that respiration in tree tissues is strongly linked to their nitrogen content [33], which may explain differences in respiration levels in forest stands [24]. Pruyn et al. [28] found that sapwood of older trees had higher respiratory potential than sapwood of younger trees if the outer-bark surface area of stems was used as a basis for comparison of respiratory potential. Araki et al. [3, 4] reported significant variations of respiration along stems of *Chamaecyparis obtusa* trees, with maximal respiration rates being observed inside the tree crown, especially during the growing season.

Based on these observations, if allometry is a general case in tree body structure, it might be supposed that the distribution of variously active surfaces within the stems of larger trees should be different from that in smaller trees. In particular, it is not improbable that the share of actively respiring surfaces is higher in larger trees, at least in some species.

Another consideration deals with the overall nonlinearity of tree growth and composition of datasets. It is widely known that a typical tree has an S-shaped growth curve, both in linear and volumetric terms. If one considers a dataset with no mature and overmatured trees then for this particular dataset larger trees will on average grow faster than small trees will. Because growth respiration is a substantial part of the total respiration the nonlinearity of the growth curve means that larger trees would show higher respiration on both an absolute and relative per area unit basis. Meir and Grace [24] showed that maintenance respiration in slower-growing climax-stage species is relatively higher. Cheng et al. [8] analyzed a multispecies dataset consisting of angiosperms and gymnosperms and concluded that in general the numeric values for scaling exponents will depend on the species pooled in the analysis and the *range of sizes within datasets*.

As a conclusion it should be admitted that these explanations will remain theoretical speculations until a crucial experiment is performed. The reader may take notice of a gap in the published data. Respiration studies do not as a rule include relationships between stem volume/mass and stem surface area. Allometric studies are mostly not aimed at measuring respiration. The crucial experiment should include measurements of both scaling exponents, volume versus surface area and respiration versus volume/mass in different sites. Such kinds of data would help to clarify the issue if the relative per unit area respiration scales isometrically with increase of surface size or deviates from isometry.

References

1. Amthor JS (1989) Respiration and crop productivity. Springer, New York
2. Amthor JS (2000) The McCree–de Wit–Penning de Vries–Thornley respiration paradigms: 30 years later. Ann Bot 86(1):1–20
3. Araki MG, Utsugi H, Kajimoto T, Han Q, Kawasaki T, Chiba Y (2010) Estimation of whole-stem respiration, incorporating vertical and seasonal variations in stem CO2 efflux rate, of Chamaecyparis obtusa trees. J For Res 15(2):115–122
4. Araki MG, Kajimoto T, Han Q, Kawasaki T, Utsugi H, Gyokusen K, Chiba Y (2015) Effect of stem radial growth on seasonal and spatial variations in stem co2 efflux of chamaecyparis obtusa. Trees 29(2):499–514
5. Beck CB (2010) An introduction to plant structure and development: plant anatomy for the twenty-first century. Cambridge University Press, Cambridge
6. Bosc A, De Grandcourt A, Loustau D (2003) Variability of stem and branch maintenance respiration in a pinus pinaster tree. Tree Physiol 23(4):227–236
7. Cheng D-L, Li T, Zhong Q-L, Wang G-X (2010) Scaling relationship between tree respiration rates and biomass. Biol Lett 6(5):715–717
8. Cheng D, Niklas KJ, Zhong Q, Yang Y, Zhang J (2014) Interspecific differences in whole-plant respiration vs. biomass scaling relationships: a case study using evergreen conifer and angiosperm tree seedlings. Am J Bot 101(4):617–623
9. Curtis RO, Marshall DD et al (2009) Levels-of-growing-stock cooperative study in Douglas-fir: Report No. 19–the Iron Creek study, 1966–2006. United States Department of Agriculture, Forest Service, Pacific Northwest Research Station
10. Damesin C, Ceschia E, Le Goff N, Ottorini J-M, Dufrene E (2002) Stem and branch respiration of beech: from tree measurements to estimations at the stand level. New Phytol 153(1):159–172
11. Evans LT (1996) Crop evolution, adaptation and yield. Cambridge University Press, Cambridge
12. García O (1990) Growth of thinned and pruned stands. In: New approaches to spacing and thinning in plantation forestry: proceedings of a IUFRO symposium, Rotorua, 10–14 April 1989. International Union of Forestry Research Organisations, pp 84–97
13. Gavrikov VL (2015) Whether respiration in trees can scale isometrically with bole surface area: a test of hypothesis. Ecol Model 312:318–321
14. Hoque ATMR, Suwa R, Shigeta M, Hagihara A (2008) Comparison of the size-dependence of aboveground respiration between Kandelia obovata and Bruguera gymnorrhiza on Okinawa Island, Japan. In: Puangchit L, Diloksumpun S (eds) FORTROP II international conference tropical forestry change in a changing world (Kasetsart University, Bangkok, November 17–20, 2008). Faculty of Forestry Kasetsart University, Bangkok, pp 55–71
15. Hoque ATMR, Sharma S, Suwa R, Mori S, Hagihara A (2010) Seasonal variation in the size-dependent respiration of mangroves Kandelia obovata. Mar Ecol Prog Ser 404:31–37
16. Hoque ATMR, Sahadew S, Suwa R, Mori S, Hagihara A (2012) Nighttime respiration behavior of Kandelia obovata Sheue, Liu and Yong in Subtropical Okinawa Island. Bangladesh J For Sci 32(1):74–78
17. Hoque ATMR, Sahadew S, Suwa R, Mori S, Hagihara A (2013) Aboveground nighttime respiration behavior of a mangrove tree Bruguiera gymnorrhiza in Subtropical Okinawa Island, Japan. Bangladesh Agric 5(1):44–48
18. Inoue A, Kurokawa Y (2001) Theoretical derivation of a two-way volume equation in coniferous species. J Jpn For Soc (Jpn) 83(2):130–134
19. Inoue A, Nishizono T (2015) Conservation rule of stem surface area: a hypothesis. Eur J For Res 134(4):599–608
20. Kinerson RS (1975) Relationships between plant surface area and respiration in loblolly pine. J Appl Ecol 12(3):965–971
21. Korzuhin MD, Tselniker YL, Semenov SM (2008) Ecophysiological model of primary production of woody plants and estimations of climatic limits of their distribution. Meteorol Gidrol 12:56–69 (in Russian)

22. Luft K, Schaefer W, Wiegleb G (1993) 50 Jahre NDIR-Gasanalyse. Tech Mess 60:363–371
23. Marshall DD, Curtis RO (2001) Levels-of-growing-stock cooperative study in Douglas-fir: report no. 15-Hoskins: 1963–1998. United States Department of Agriculture, Forest Service
24. Meir P, Grace J (2002) Scaling relationships for woody tissue respiration in two tropical rain forests. Plant Cell Environ 25(8):963–973
25. Mohr H, Schopfer P (1995) Plant physiology. Springer, Berlin/Heidelberg/New York
26. Mori S, Yamaji K, Ishida A, Prokushkin SG, Masyagina OV, Hagihara A, Hoque AR, Suwa R, Osawa A, Nishizono T et al (2010) Mixed-power scaling of whole-plant respiration from seedlings to giant trees. Proc Natl Acad Sci 107(4):1447–1451
27. Niklas KJ (2004) Plant allometry: is there a grand unifying theory? Biol Rev 79(04):871–889
28. Pruyn ML, Gartner BL, Harmon ME (2002) Respiratory potential in sapwood of old versus young ponderosa pine trees in the Pacific Northwest. Tree Physiol 22(2–3):105–116
29. Pruyn ML, Gartner BL, Harmon ME (2005) Storage versus substrate limitation to bole respiratory potential in two coniferous tree species of contrasting sapwood width. J Exp Bot 56(420):2637–2649
30. Raven PH, Evert RF, Eichhorn SE (1992) Biology of plants, 5th edn. Worth Publishers, New York
31. Rayment MB, Loustau D, Jarvis PJ (2002) Photosynthesis and respiration of black spruce at three organizational scales: shoot, branch and canopy. Tree Physiol 22(4):219–229
32. Reich PB, Tjoelker MG, Machado J-L, Oleksyn J (2006) Universal scaling of respiratory metabolism, size and nitrogen in plants. Nature 439(7075):457–461
33. Reich PB, Tjoelker MG, Pregitzer KS, Wright IJ, Oleksyn J, Machado J-L (2008) Scaling of respiration to nitrogen in leaves, stems and roots of higher land plants. Ecol Lett 11(8):793–801
34. Romanovskii MG (2014) Polythene and hystogenesis in forest plants. Printing house "Cyrillic", Nizhny Novgorod (in Russian)
35. Ryan MG, Waring RH (1992) Maintenance respiration and stand development in a subalpine lodgepole pine forest. Ecology 73(6):2100–2108
36. Ryan MG, Linder S, Vose JM, Hubbard RM (1994) Dark respiration of pines. Ecol Bull 43:50–63
37. Ryan MG, Gower ST, Hubbard RM, Waring RH, Gholz HL, Cropper WP Jr, Running SW (1995) Woody tissue maintenance respiration of four conifers in contrasting climates. Oecologia 101(2):133–140
38. Soloviev VA (1983) Respiration gas exchange in wood. Izd-vo Leningradskogo Universiteta, Leningrad (in Russian)
39. Sprugel DG (1990) Components of woody-tissue respiration in young Abies amabilis (Dougl.) Forbes trees. Trees 4(2):88–98
40. Sprugel DG, Ryan MG, Brooks J, Vogt K, Martin TA (1995) Respiration from the organ level to the stand. In: Smith WK, Hinkley TM (eds) Resource physiology of conifers. Academic Press, San Diego, pp 255–299
41. West GB, Brown JH, Enquist BJ (1997) A general model for the origin of allometric scaling laws in biology. Science 276(5309):122–126
42. West GB, Brown JH, Enquist BJ (1999) A general model for the structure and allometry of plant vascular systems. Nature 400(6745):664–667
43. Wilson BF (1966) Mitotic activity in the cambial zone of pinus strobus. Am J Bot 53(4):364–372

Conclusion

Decades of research in botany, physiology, forest science, and other disciplines have shown that trees are quite a peculiar, interesting, and extremely important branch of the plant kingdom.

Trees are best understood through how secondary growth is implemented in them. Secondary growth is a widespread feature among higher plants but trees developed a certain version of it producing what we call a tree-like growth habit. Particularly, secondary growth is architecturally organized in trees in such a way that growing tissue forms an unbroken sheath that covers stems, all branches, and coarse roots except fine roots and tips of shoots where buds perform primary growth.

One of the pillars of classical biology is understanding that organisms are large collections of cells. Eventually, growth that we see macroscopically is the division and growth of cells. In a sense, an organism is its cells. Therefore one might hope that the best way to describe organisms is to follow the growth of individual cells in it. Leaving aside unicellular organisms, even for the smallest multicellular beings such a hope is a fantasy rather than a reality because of the exponential outbreak of cell quantity at the very initial stages of ontogeny, not to mention a growing variety of cell specializations.

The cell-focused view has therefore been successfully challenged many times by a functional approach focusing not on how many cells there are but on what they do in the sense of the substance and energy flows they produce. The functional approach has been reflected in much research denoted as process-oriented modeling. Functional modeling has been successful in many respects but recent opinions pointed out that it was still necessary to look into the cell level in the search for growth rate limitations. These limitations come from how fast real cells can divide and enlarge their volume.

In order to avoid an explicit consideration of cells their morphological representative may be used in modeling. Because architecturally living tissues of trees

© The Author(s) 2017
V.L. Gavrikov, *Stem Surface Area in Modeling of Forest Stands*,
SpringerBriefs in Plant Science, DOI 10.1007/978-3-319-52449-8

are surfaces, surface measures are good candidates to be representatives. In respect to photosynthesizing tissues, it is standard to use measures such as total leaf area. Regarding the tissues of secondary growth, a natural choice may be the stem surface area as a proxy of cambial sheath.

In contrast to linear or even volumetric dimensions, surfaces of real organisms are hard to evaluate because of the many irregularities in their shapes. In the case of trees, their large sizes also do not make the surface measurement easier. On the other hand, there is not much need for data on the stem surface area in industry. The standing volume of a commercial forest stand is by definition a volume. Trade with wood is in cubic units. Stem diameter and height are more important data for industrial use. Hopes to relate the stem surface area to increment assessments have not yet come true, at least in mainstream forestry. Probably, the difficulties of measurement coupled with the vague immediate usefulness led to interest in stem surface area being rather unstable in forest science.

For research purposes, however, some successful approaches have been developed to estimate the stem surface area and to use it in studies of forest stand development. The most straightforward way is to use a combination of linear dimensions and relate it to "genuine" stem surface area through regression. If the received coefficients are more or less stable then regression may be used to find the stem surface area of other forest stands. Therefore in the literature a simple product of mean diameter at breast height (dbh) and mean tree height multiplied by a coefficient was suggested to represent the total stem surface area of forest stand. It is easy to see that this approach implies a conic approximation of a tree's stem shape because a cone surface area is the product of radius and generatrix multiplied by known coefficients. The radius is just halved diameter and the generatrix value is very close to height for such narrow shapes as trees.

Trees are of course not cones but a conic approximation does not necessarily require them to be cones. It requires rather a sort of proportionality to simple conic shape, which naturally may be a source of deviations. At the expense of some loss of accuracy, a conic approximation provides a very convenient and transparent mathematical form that can be consistently used in modeling. In this book, a geometrical model of an even-aged/even-sized forest was formulated in which a conic mathematical representation of the total stem surface area was a key point that provided an opportunity for analytical study. The main focus of the study was on subtle internal relations that are not readily obvious in the structure of a forest stand.

By measuring various forest stand dimensions, whether in a number of different stands or in one stand consecutively over time, one gets data allowing the relation of the dimensions to each other. An incomplete list of the relations may, for example, include "how stand volume depends on age," "how stand density depends on age," "how mean tree sizes depend on stand density," and "how a linear dimension depends on another linear dimension," among others. These relationships may be called primary relationships because they are built on primarily measured dimensions.

A dense forest stand is not an occasional collection of independent trees but a system in which trees intensively compete, which is reflected in their growth rates, sizes, and mortality. If so, a question may arise whether the primary relationships reflecting links between the growth rates, sizes, and mortality are also not independent of each other. What if there are also secondary relationships, a kind of "relationships between relationships"?

In this book, three relationships were explored: first, how total stem surface area depends on stand density; second, how stem height depends on dbh (represented by stem radius); and third, how dbh depends on stand density. It has been theoretically found that all these relationships were tightly interrelated. Especially interesting, if it is known that total stem surface area is a constant, independent of stand density, then exponent parameters in the second and the third relationships can be directly computed from one another. It means that there is a relationship between the second and the third relationships. Testing of the model formulas against real forest stand data showed that in spite of extreme simplicity the model *quantitatively* well predicts relations between parameters (exponents) of different relationships.

Based on total stem surface area consideration, another application of the model was about the famous $-3/2$ self-thinning rule. Much debate in the literature has touched on various aspects of the rule, initially enthusiastic and then more and more critical. Much research has shown that the slope of the self-thinning curve was not constant in deviating from $-3/2$ as growth of a forest stand progressed. Moreover, evidence appeared that the slope was not universal but rather species-specific. It was shown in this book that modeling the total stem surface area may help to disentangle many controversies about the self-thinning rule.

There is a peculiar time span in the life of a forest stand undergoing self-thinning. That time is when total stem surface area stays constant independent of continuing self-thinning, that is, loss of dying-off stems. Analysis of field data with the help of the model has shown that it was the very period of time when the slope of the self-thinning line reached the value of $-3/2$. Before this period of time, the slope deviated from $-3/2$ and in the course of further ongoing self-thinning the slope again deviated from $-3/2$. It is important therefore to note that the model considered self-thinning in a broader context and, in a sense, helped the $-3/2$ self-thinning rule to find its deserved place in the whole picture of the self-thinning process.

In the course of forest stand development, the total stem surface area consistently undergoes phases from increase to decline. In between, however, its value may stop changing and remain relatively constant while the number of stems continues to fall. Hypothetically, this phase is a subtle balance of growth and mortality when they compensate each other. Another probable view is that the phase reflects the fact that the particular site cannot bear more living cells than a definite maximal level, which resembles processes in the stand canopy where the leaf area may stop growing after canopy closure. In any case, it seems that the dynamics of the total stem surface area may present an interesting and promising research topic.

Appendix

See Tables A.1, A.2, A.3, A.4, A.5, A.6, A.7, A.8, A.9, A.10, A.11, A.12, A.13, A.14, A.15, A.16, A.17, A.18, A.19, A.20, and A.21.

Table A.1 Fragment of control plot data of levels-of-growing-stock cooperative study in Douglas fir (Hoskins experiment) published by Marshall and Curtis [4]

Age, year	Mean height, m	DBH, cm	QMD, cm	Density, trees/ha	Volume stock, m^3
20	13.4	19.2	9.8	4244	140.2
23	16.1	23.0	11.6	4043	237.9
27	20.3	27.6	14.7	3141	379.2
30	23.2	30.5	16.8	2684	485.7
32	25.0	32.2	18.3	2318	533.3
36	27.8	35.1	20.8	1894	650.5
40	30.7	38.1	23.2	1614	763.4
45	34.2	41.5	26.9	1206	851.8
50	37.8	44.7	30.7	930	936.9
55	40.2	48.3	34.2	757	1002.2

Mean height = mean height of 100 tallest trees, DBH = diameter at breast height, QMD = quadratic mean diameter

Table A.2 Fragment of control plot data of levels-of-growing-stock cooperative study in Douglas fir (Iron Creek experiment) published by Curtis and Marshall [1]

Age, year	Mean height, m	DBH, cm	QMD, cm	Density, trees/ha	Volume stock, m³
19	11.1	16.6	8.9	3153	81.6
23	14.8	21.0	11.2	3092	167.1
26	17.4	24.2	12.9	2985	250.3
30	20.6	27.4	14.8	2775	365.9
33	23.2	29.7	16.3	2507	450.4
37	26.1	32.2	17.9	2239	546.5
42	29.8	35.0	20.1	1865	650.5
47	32.9	38.3	22.7	1453	704.6
52	36.0	41.3	25.5	1231	814.1
59	40.1	45.4	28.9	1017	957.8

Mean height = mean height of 100 tallest trees, DBH = diameter at breast height, QMD = quadratic mean diameter

Table A.3 Fragment of control plot data of levels-of-growing-stock cooperative study in Douglas fir (Rocky Brook experiment) published by Curtis and Marshall [2]

Age, year	Mean height, m	DBH, cm	QMD, cm	Density, trees/ha	Volume stock, m³
29	11.9	18.2	9.5	3491	99.0
33	14.0	20.4	10.5	3446	139.3
40	17.7	25.2	12.8	3227	244.2
46	21.0	28.5	14.7	2935	349.8
52	23.6	31.4	16.5	2614	442.0
59	26.1	34.6	18.7	2231	536.7
70	29.7	38.2	22.0	1643	615.4

Mean height = mean height of 100 tallest trees, DBH = diameter at breast height, QMD = quadratic mean diameter

Table A.4 Fragment of control plot data of levels-of-growing-stock cooperative study in Douglas fir (Skykomish experiment) published by King et al. [3]

Age, year	Mean height, m	DBH, cm	QMD, cm	Density, trees/ha	Volume stock, m³
24	13.8	20.5	12.1	1467	88.6
28	17.5	25.1	14.8	1467	172.9
31	20.5	28.3	16.9	1454	258.9
34	23.3	31.1	18.6	1436	353.4
38	27.1	35.0	21.1	1300	463.7
42	30.8	38.8	23.8	1161	584.4
46	34.0	42.7	26.2	1047	724.4
51	37.3	46.8	29.0	951	847.9
56	40.7	50.4	32.9	750	931.8

Mean height = mean height of 100 tallest trees, DBH = diameter at breast height, QMD = quadratic mean diameter

Table A.5 Fragment of control plot data of levels-of-growing-stock cooperative study in Douglas fir (Clemons experiment) published by King et al. [3]

Age, year	Mean height, m	DBH, cm	QMD, cm	Density, trees/ha	Volume stock, m³
19	11.9	19.5	10.2	1696	59.6
22	14.9	23.1	12.5	1688	111.6
26	18.5	27.2	15.0	1634	193.1
29	21.3	30.0	16.7	1589	269.7
32	23.8	32.5	18.3	1498	337.7
36	26.1	35.1	20.1	1379	411.0
40	28.8	37.4	21.8	1235	477.2
45	32.0	40.3	24.4	1070	568.7
50	34.7	43.3	26.8	955	659.4

Mean height = mean height of 100 tallest trees, DBH = diameter at breast height, QMD = quadratic mean diameter

Table A.6 Mironenko-98 dataset[a] [5, p. 239]

Age, years	Stand density, trees/ha	DBH cm	Mean height, m	Volume stock, m³/ha
100	465	32.6	30.1	515
110	426	34.1	31.1	540
120	390	36.0	32.1	566
130	363	37.7	33.0	590
140	338	39.4	33.8	612
150	309	41.5	34.5	632
70	702	25.1	24.7	402
80	613	27.3	26.2	435
90	548	29.3	27.5	465
100	490	31.4	28.8	495
110	445	33.3	29.9	522
120	407	35.2	31.0	549
130	376	37.0	32.0	574
140	348	38.8	32.9	598
150	325	40.5	32.8	621

Pine cultures, Tambov region, Russian Federation; site quality I
[a]Data in Tables A.6, A.7, A.8, A.9, A.10, A.11, A.12, A.13, A.14, A.15, A.16, A.17, and A.18 were first published by Usoltsev [5]

Table A.7 Kozhevnikov-84 dataset [5, p. 31]

Age, years	Stand density, trees/ha	DBH cm	Mean height, m	Volume stock, m³/ha
60	1360	19.8	19.8	387
40	2340	14.1	15.6	281
30	3460	10.9	12.5	204
20	5630	7.2	8.4	107
15	7510	5.1	6.0	56

Pine cultures, Belorussia; site quality I

Table A.8 Gruk-79 dataset [5, p. 30]

Age, years	Stand density, trees/ha	DBH cm	Mean height, m	Volume stock, m³/ha
10	7274	4.9	4.3	44
15	6059	6.4	6.6	83
20	5667	7.3	8.2	116
25	4809	8.5	9.8	151
30	3792	9.9	12.4	186
35	3410	10.7	14.0	219
40	2449	12.8	15.7	258

Pine cultures, Belorussia; site quality I

Table A.9 Uspensky-87 dataset [5, p. 240]

Age, years	Stand density, trees/ha	DBH cm	Mean height, m	Volume stock, m³/ha
120	171	44.8	32.7	374
120	201	41.2	30.9	355
120	238	37.7	29.1	335
100	240	37.8	29.1	335
100	283	34.5	27.4	317
80	354	30.7	25.2	293
80	432	27.7	23.4	274
60	513	25.3	21.9	257
60	650	22.3	19.9	235
40	954	18.0	16.8	199
40	1244	15.6	14.9	177
30	1533	13.8	13.3	157
30	1931	12.0	11.7	137
20	2758	9.5	9.2	105
20	3218	8.4	8.0	88
10	4198	5.9	4.9	44
10	4240	5.4	4.2	34

Pine cultures, Tambov region, Russian Federation; site quality I

Table A.10 Gabeev-90 dataset [5, p. 482]

Age, years	Stand density, trees/ha	DBH cm	Mean height, m	Volume stock, m³/ha
40	2137	15.2	16.7	298
45	1546	16.6	17.9	306
45	2323	16	17.9	338
83	543	36.8	26.4	665
86	430	37	26.4	522

Pine natural forests, Novosibirsk region, Russian Federation; site quality I

Table A.11 Heinsdorf and Krauß-90 dataset [5, p. 56]

Age, years	Stand density, trees/ha	DBH cm	Mean height, m	Volume stock, m³/ha
50	1385	18.8	18.2	329
60	942	23.3	20.8	383
70	687	27.6	23.0	426
80	528	31.6	24.9	461
90	423	35.4	26.5	488
100	349	38.9	27.9	508
110	297	42.0	29.0	520
120	258	44.8	29.9	528

Pine natural forests, Eberswalde, Germany; site quality I

Table A.12 Kurbanov-02 dataset [5, p. 211]

Age, years	Stand density, trees/ha	DBH cm	Mean height, m	Volume stock, m³/ha
76	643	27.3	24.3	420
125	268	38.9	33.2	510
84	925	24.2	25.4	400
78	745	24.7	25.0	390
128	259	39.5	33.9	490
80	672	29.3	24.8	420

Pine natural forests, Yorshkar-Ola, Russian Federation; site quality I

Table A.13 Lebkov and Kaplina-97 dataset [5, p. 203]

Age, years	Stand density, trees/ha	DBH cm	Mean height, m	Volume stock, m³/ha
25	3287	9.9	11.2	163
51	1429	18.6	19.9	386
77	687	26.0	26.0	428
25	4331	9.1	12.4	199
47	1667	16.5	18.2	332
58	1263	19.6	19.5	354
60	1040	21.1	22.5	400

Pine natural forests, Vladimir region, Russian Federation; site quality I

Table A.14 Uspensky-87 dataset [5, p. 240]

Age, years	Stand density, trees/ha	DBH cm	Mean height, m	Volume stock, m³/ha
30	1078	16.9	15.9	189
30	1271	15.4	14.7	175
40	656	22.2	19.8	234
40	796	20.0	18.3	217
60	348	31.0	25.4	295
60	417	28.2	23.7	277
80	246	37.1	28.8	332
80	294	33.8	27.0	312
100	199	41.4	31.0	356

Pine cultures, Tambov region, Russian Federation; site quality Ia

Table A.15 Uspensky-87 dataset [5, p. 240]

Age, years	Stand density, trees/ha	DBH cm	Mean height, m	Volume stock, m³/ha
80	549	24.4	21.3	250
100	333	31.7	25.8	300
100	429	27.8	23.6	276
120	288	34.2	27.2	315

Pine cultures, Tambov region, Russian Federation; site quality II

Table A.16 Mironenko-98 dataset [5, p. 239]

Age, years	Stand density, trees/ha	DBH cm	Mean height, m	Volume stock, m³/ha
50	889	21.4	22.8	348
60	754	23.8	24.6	387
70	653	26.1	26.2	423
80	577	28.2	27.6	455
90	515	30.3	28.9	485
50	960	20.5	21.2	327
60	748	23.8	23.0	365

Pine cultures, Tambov region, Russian Federation; site quality Ia

Table A.17 Heinsdorf and Krauß-90 dataset [5, p. 56]

Age, years	Stand density, trees/ha	DBH cm	Mean height, m	Volume stock, m³/ha
40	1751	17.0	17.8	332
50	1100	22.0	21.1	408
60	759	27.0	24.0	472
70	559	31.9	26.5	522
80	434	36.3	28.6	563
90	351	40.4	30.4	594
100	292	44.2	31.9	616
110	250	47.6	33.1	631
120	218	50.7	34.2	638

Pine natural forests, Eberswalde, Germany; site quality Ia

Table A.18 Heinsdorf and Krauß-90 dataset [5, p. 56]

Age, years	Stand density, trees/ha	DBH cm	Mean height, m	Volume stock, m³/ha
25	9399	6.0	8.2	98
30	5924	8.0	9.5	146
40	3063	11.7	12.6	206
50	1838	15.6	15.2	258
60	1223	19.5	17.5	302
70	878	23.3	19.4	339
80	665	27.0	21.2	368
90	527	30.4	22.7	391
100	431	33.6	23.9	408
110	363	36.3	24.9	420
120	313	39.1	25.8	426

Pine natural forests, Eberswalde, Germany; site quality II

Table A.19 Data of a Scots pine stem analysis

H, m	D, cm	Age, year						
		227	220	210	200	190	180	170
0	D_{NS}	60.7	59.9	58.5	56.3	54.4	52.3	50.6
	D_{EW}	57.9	57.3	56.0	54.2	52.6	51.0	50.1
1	D_{NS}	53.4	52.7	51.5	49.9	48.4	46.8	45.4
	D_{EW}	53.3	52.4	51.3	49.7	48.0	46.4	45.3
3	D_{NS}	52.2	51.8	51.0	49.6	48.3	46.3	45.4
	D_{EW}	46.2	45.6	44.5	43.2	41.7	40.4	39.5
5	D_{NS}	51.7	51.0	50.1	48.9	47.5	45.9	44.7
	D_{EW}	46.0	45.2	44.1	42.7	41.4	40.0	38.9
7	D_{NS}	49.4	48.9	47.9	46.6	45.3	43.9	42.6
	D_{EW}	43.2	42.5	41.4	40.1	38.9	37.6	36.4
9	D_{NS}	47.8	47.0	45.9	44.4	42.9	41.4	40.0
	D_{EW}	42.5	41.9	40.8	39.5	38.3	37.0	35.9
11	D_{NS}	45.2	44.5	43.5	42.3	41.0	39.8	38.8
	D_{EW}	41.4	40.6	39.5	38.4	37.1	35.8	34.7
13	D_{NS}	44.7	44.0	42.9	41.7	40.4	39.0	37.8
	D_{EW}	40.3	39.4	38.1	36.7	35.3	34.1	33.1
15	D_{NS}	40.1	39.5	38.3	37.1	35.7	34.3	33.2
	D_{EW}	42.5	41.8	40.4	39.0	37.5	36.0	34.6
17	D_{NS}	39.1	38.1	36.7	35.2	33.5	32.1	30.8
	D_{EW}	35.4	35.8	34.6	33.4	32.0	30.7	29.3
19	D_{NS}	35.9	34.7	33.3	31.9	30.3	28.9	27.5
	D_{EW}	35.8	34.9	33.6	32.0	30.6	29.1	27.8
21	D_{NS}	30.0	29.2	27.7	26.4	24.7	23.2	21.5
	D_{EW}	31.0	30.0	28.6	26.9	25.2	23.6	22.1
23	D_{NS}	23.7	22.7	21.2	19.5	17.7	16.7	14.6
	D_{EW}	21.8	21.0	19.6	18.2	16.7	15.4	13.9
25	D_{NS}	13.0	12.3	11.0	9.5	8.2	7.1	5.8
	D_{EW}	12.2	11.6	10.7	9.6	8.1	6.8	5.6
26	D_{NS}	9.0	8.3	7.3	5.8	4.2	3.1	2.0
	D_{EW}	8.5	7.8	6.8	5.6	4.2	3.0	2.0
	H_{stem}, m	28.07	27.9	27.6	27.4	27.1	26.8	26.6
	H_{top}, m	2.07	1.9	1.6	1.4	1.1	0.8	0.6
	D_{base}, cm	8.70	8.0	7.0	5.7	4.2	3.0	2.0

H = height at which a cross-sectional cut is taken, D = diameter, D_{NS} = diameter in the north-south direction, D_{EW} = diameter in the east-west direction, H_{stem} = height of stem in the corresponding age without the top, H_{top} = height of the tree top, and D_{base} = diameter of the top base. The data on stem analysis was provided by V.V. Kuzmichev

Table A.20 Data of a Scots pine stem analysis (continued from Table A.19)

H, m	D, cm	Age, year							
		160	150	140	130	120	110	100	90
0	D_{NS}	48.8	46.3	44.3	41.9	39.4	36.5	33.8	31.1
	D_{EW}	48.3	46.0	43.8	41.1	38.5	35.6	33.1	30.4
1	D_{NS}	43.9	42.0	40.4	38.1	35.7	33.1	30.8	28.4
	D_{EW}	43.7	41.6	39.8	37.6	35.2	32.8	30.6	29.2
3	D_{NS}	43.8	41.7	39.5	37.3	35.0	32.7	30.5	28.1
	D_{EW}	38.4	37.0	35.7	34.0	32.0	30.4	28.7	26.9
5	D_{NS}	43.2	41.3	40.3	38.0	35.9	33.8	31.5	29.3
	D_{EW}	37.7	36.5	35.4	34.0	32.2	30.4	28.5	26.6
7	D_{NS}	41.0	39.2	37.3	35.2	33.1	30.9	28.7	26.5
	D_{EW}	35.2	33.8	32.7	31.3	29.5	27.8	26.2	24.6
9	D_{NS}	38.3	36.6	35.0	33.2	31.1	29.1	27.0	24.7
	D_{EW}	34.8	33.3	32.2	30.6	28.8	27.2	25.7	23.8
11	D_{NS}	37.0	35.3	33.4	31.4	29.4	27.6	25.6	23.4
	D_{EW}	33.4	32.0	30.7	29.2	27.4	25.5	23.7	22.0
13	D_{NS}	36.0	34.3	32.7	30.4	28.6	26.2	24.2	22.1
	D_{EW}	31.8	30.4	29.3	27.8	26.2	24.4	20.5	20.4
15	D_{NS}	31.7	30.1	28.6	26.8	24.9	22.9	20.9	18.8
	D_{EW}	32.4	31.3	29.6	27.8	26.0	23.8	22.0	19.3
17	D_{NS}	29.1	27.3	25.9	23.8	22.2	20.1	18.2	15.7
	D_{EW}	28.0	26.6	25.3	23.8	22.1	20.9	18.4	16.0
19	D_{NS}	26.0	24.4	22.9	21.0	19.0	16.8	15.0	12.4
	D_{EW}	26.3	24.5	23.0	21.2	19.1	17.0	15.0	12.1
21	D_{NS}	19.7	18.1	16.7	15.0	13.4	11.6	9.7	7.0
	D_{EW}	20.0	18.3	16.7	14.9	13.3	11.4	9.6	7.3
23	D_{NS}	12.9	11.2	9.6	8.2	6.8	5.5	3.4	2.0
	D_{EW}	12.2	10.6	9.3	7.7	6.3	4.6	3.3	2.0
25	D_{NS}	4.2	2.9	1.9	1.1				
	D_{EW}	4.1	2.9	2.0	1.1				
26	D_{NS}	1.1	0.4						
	D_{EW}	1.1	0.4						
	H_{stem}, m	26.3	26.1	25.6	25.2	24.9	24.5	24.1	23.8
	H_{top}, m	0.3	0.1	1.6	1.2	2.9	2.5	2.1	1.8
	D_{base}, cm	1.1	0.4	5.6	4.5	9.9	8.2	6.4	4.5

Table A.21 Data of a Scots pine stem analysis (continued from Table A.20)

H, m	D, cm	Age,year							
		80	70	60	50	40	30	20	10
0	D_{NS}	29.8	28.3	25.8	22.5	18.4	14.6	10.0	4.6
	D_{EW}	29.0	27.5	25.1	22.4	18.5	14.7	9.9	4.4
1	D_{NS}	27.1	25.8	23.5	20.8	17.1	13.4	8.8	3.3
	D_{EW}	26.9	25.4	23.1	20.6	17.2	13.5	8.7	3.3
3	D_{NS}	26.5	24.8	22.4	19.9	16.1	11.9	5.7	
	D_{EW}	25.5	24.0	21.8	19.0	15.4	11.2	5.2	
5	D_{NS}	27.5	26.3	24.2	21.3	17.8	13.7	4.6	
	D_{EW}	25.3	23.8	21.4	18.7	14.8	10.6	4.1	
7	D_{NS}	25.1	23.5	20.8	17.9	13.9	8.9		
	D_{EW}	23.4	21.9	19.5	16.8	13.1	8.3		
9	D_{NS}	23.2	21.4	18.8	15.9	11.6	5.8		
	D_{EW}	22.3	20.8	18.1	15.1	11.2	5.7		
11	D_{NS}	21.8	19.8	16.9	13.9	9.4	3.3		
	D_{EW}	20.6	19.2	16.8	13.6	9.3	3.1		
13	D_{NS}	20.1	18.4	15.4	11.7	7.3			
	D_{EW}	18.9	17.2	14.6	11.4	7.1			
15	D_{NS}	16.5	14.7	12.0	8.2	3.9			
	D_{EW}	17.5	15.5	12.2	8.3	4.0			
17	D_{NS}	13.5	11.7	8.6	4.8				
	D_{EW}	13.5	11.7	8.6	4.9				
19	D_{NS}	9.8	7.5	4.5					
	D_{EW}	9.6	7.3	4.4					
21	D_{NS}	5.7	3.0						
	D_{EW}	5.0	2.8						
23	D_{NS}	0.9							
	D_{EW}	0.9							
25	D_{NS}								
	D_{EW}								
26	D_{NS}								
	D_{EW}								
	H_{stem},m	23.4	22.8	20.8	18.7	16.6	12.8	6.6	1.7
	H_{top},m	1.4	2.8	2.8	2.7	2.6	2.8	2.6	1.7
	D_{base},cm	3.1	5.1	6.5	6.5	5.5	4.4	4.8	4.5

References

1. Curtis RO, Marshall DD et al (2009) Levels-of-growing-stock cooperative study in Douglas-fir: Report No. 19–the Iron Creek study, 1966–2006. United States Department of Agriculture, Forest Service, Pacific Northwest Research Station
2. Curtis RO, Marshall DD et al (2009) Levels-of-growing-stock cooperative study in Douglas-fir: Report No. 18-Rocky Brook, 1963–2006. US Department of Agriculture, Forest Service, Pacific Northwest Research Station
3. King JE, Marshall DD, Bell JF (2002) Levels-of-growing-stock cooperative study in Douglas-fir: Report No. 17-the Skykomish study, 1961–93; the Clemons study, 1963–1994. Pacific Northwest Research Station, USDA Forest Service
4. Marshall DD, Curtis RO (2001) Levels-of-growing-stock cooperative study in Douglas-fir: Report No. 15-Hoskins: 1963–1998. United States Department of Agriculture, Forest Service
5. Usoltsev VA (2010) Eurasian forest biomass and primary production data. Ural Branch of Russian Academy of Sciences, Yekaterinburg (in Russian)

Index

Printed in the United States
By Bookmasters